Advances in Historical Ecology

Advances in Historical Ecology

Jaspal Juneja

CYBER TECH PUBLICATIONS
4264/3, Ansari Road, Daryaganj, New Delhi-110002 (India)
Ph.: 011-23244078. Fax: 011-23280028
E-Mail: cyberpublicationsdelhi@yahoo.com

ADVANCES IN HISTORICAL ECOLOGY

Jaspal Juneja

First Edition 2009

Published by :

G.S. Rawat for **Cyber Tech Publications**
4264/3, Ansari Road, Daryaganj, New Delhi-110002 (India)
Ph.: 011-23244078. Fax: 011-23280028
E-Mail: cyberpublicationsdelhi@yahoo.com

Preface

The study of ecology is an attempt to understand the reciprocal relationships between the living and nonliving elements of the earth. For years, however, the discipline has either neglected the human element entirely or presumed that its effect on natural ecosystems is invariably negative. However, social scientists in geography and anthropology have criticized efforts to address this human-environment interaction as deterministic and mechanistic.

Bridging the gap between social and natural sciences, the undersigned has used a more holistic perspective to explore the relationships between humans and their environment. Exploring short - and long-term local and global change, eighteen specialists in anthropology, geography, history, ethnobiology, and related disciplines present new perspectives on historical ecology.

A broad theoretical background of the material factors central to the field, such as anthropogenic fire, soils, and pathogens is being presented. This knowledge, applied to a series of regions, is used to investigate landscape transformations over time in South America, the Mississippi Delta, the Great Basin, Thailand and India. The contributors focus on traditional societies where lands are most at risk from the incursions of complex, state-level societies.

The book lays the groundwork for a more meaningful understanding of humankind's interaction with its biosphere. Scholars and environmental policymakers alike will appreciate this new critical vocabulary for biocultural phenomena. I will feel happy if readers give their feedback for the same.

—Author

Preface

CONTENTS

1

The Historical Ecology of Global Climate Change

INTRODUCTION

The assumption that "culture has triumphed over nature," is mistaken, and characterizes an outdated nature-culture dualism. While in Anthropological human evolution textbooks the first part of the story is couched in evolutionary and environmental terms, the second part denies the environment a meaningful role in human history. Instead values, beliefs and issues, history, and culture constitute the key elements of the explanatory framework. This also reflected in the disciplinary separation of archeologist/ physical anthropologists versus sociocultural anthropologists: neither acknowledges their mutual reliance.

Few efforts have been made that incorporate information about how humans have altered the environment or about how environmental change revised human activity. Examples of such changes are subsistence strategies, demographical patterns, and perceptions. To achieve this, there exists a need to develop a multidisciplinary framework. Multidisciplinarity in science is, and has been difficult to establish (Snow). Anthropology plays an important role in the development of such an framework. Its current perspective is integrative and comparative; inclusive of temporal, spatial and cultural dimensions; and dynamic. It motivates an historical focus on the dynamics of change.

ECOLOGY

Ecology is the study of the "relationships among living organisms or between them and the physical environment."

Some characteristics of the scientific study of ecology:

- Macro-scale ecologists (global-scale ecology, '60's) have lost political ground to micro-scale ecologists.
- Landscape ecology recognizes human influence on non-human species, yet persist in distinguishing human from natural landscapes.
- Human ecology concentrates on systematic and evolutionary aspects, while social ecology emphasizes behavior. Both study human-environmenal relationships in distant past or present. Sociobiology is a paradigm within human social ecology
- Cultural ecology and cultural geography examine adaptive strategies. Both are cognizant of the role of culture in human adaptation, but not interested in long term change.
- Environmental History is the intellectual history of the environmental movement, including the political and economical implications of environmental interaction.
- Environmental Ethics explores value systems as they relate to human conduct.

None of these fields has truly integrative approach, and many lack an explicit historical component. What is needed is a multi-scalar temporal and spatial frame with an explicit focus on the role of human cognition in the human-environmental dialectic.

HISTORICAL ANALOGS

Global climate change is one of the most pressing event of current times. The anticipated changes demand investigations into patterns of human adaptation to climatic variability and change. However, the global climate change models used by physical scientists to predict climatic changes do not discriminate among biotic zones or anywhere near a human scale. Furthermore, many physical scientists assume that "novel circumstances" render any historical analogy to current anticipated global climatic change irrelevant. This attitude is due to:

(1) The lack of high quality long term (>100 yr.) instrumentally obtained data

(2) Local proxy data (such as tree ring) are only valid at the broadest temporal scales.

(3) Dismay of the comparative messiness of soft social science data

(4) Vested interest in favor of novel technologies and undervalue of traditional solutions

A regional approach could overcomes this. A region's air mass data, hydrology, soil, topography and species distribution can be used in regional models. Regionally documented ethnography, archeology, and documentary evidence evidences results of human activities and past choices which encompass the entire system. Multiple regional environmental changes can identify sensitive geographical locations. Interregional relationships may then be established and integrated with global data. This approach fosters creativity and the development of local and regional answers to global situations in which sensitive cultural issues play an important part.

HISTORICAL ANALOGS AND LANDSCAPES

Two types of historical analogs can be made:

(1) purely environmental: the global effects of the volcanic eruption of El Chichon in 1982 was similar to Krakatoa in 1883

(2) environmental and human interactions in different time periods

To study the dialectical human-environmental relationships, interactive long term sequences may be traced through the study of changing landscapes. Landscape ecology is the study of structure, function, and change of a heterogeneous land area composed of interacting ecosystems. Historical ecology or landscape history is study of past ecosystems by charting the change in landscapes over time. Thus, evidence for the historical interrelatedness of humans and environments may be read in the landscape. By interference, changing human attitudes may also be identified and their effects studied. For example, the existence of a forest is the result of both location-which determines temperature and rainfall patterns-and previous and current human management practices.

The introduction of historically informed environmental analyses into regional studies offers an important opportunity for anthropologists, archeologists, historians, and geographers. Archeology is multidisciplinary in nature (natural/physical sciences + humanities) and temporal and spatial breath required for long term analyses. Regional archeology has gone beyond the individual site, seeking to understand distribution, population and economies. Ethnohistorians are anthropologists who critically examine documents for evidence of human actions, relations and attitudes. This includes written (diaries, government documents), oral (stories about storms or pest invasions), and visual (dated drawings) documents. Enthographers study customs based on observations and understandings that guide indigenous peoples' adaptive strategies. This cultural information is transferred in complex ritual behavior or casual conversation.

Historical Ecology is the practice of a globally relevant archeology, ethnohistory, ethnography and related disciplines. While geographic information systems can give practical integration of spatial structures (habitations, soils, river drainage), practical understanding of past and current relationships among these environmental and human systems require a culturally specific temporal and spatial perspective applied at a regional scale.

An Historical Ecological Approach to Global Climate Change Expects to Identify

- The duration and frequency of air mass patterns that characterizes earlier warm episodes (bv. middle Holocene, Roman Climatic Optimum, Medieval Climatic Optimum)
- What regions where affected?
- How did biotic and human communities respond to these conditions?
- What spatial patterns are characteristic of biotic communities (species diversity and distribution) and human communities (settlement and land-use, population and agglomeration) in the earlier warm episodes?
- How might patterns be mapped in advance of contemporary global warming?

- What measures could contemporary societies employ to cope with supra-annual cycles?
- What human behaviors, attitudes, beliefs, economic strategies, and forms of governance are associated with periods of stable or unstable climate?
- What measures could contemporary societies employ to cope with a marked increase in climatic instability?
- What values ensure the greatest flexibility in adapting to changed or unstable environment. ?
- Can environmental flexibility be taught?
- How can a global environmental ethic also be culturally sensitive?
- Are some governmental structures better able than others to employ the necessary strategies of adaptation?

Integrative Themes and Considerations

Practices are maintained or modified, decisions are made, and ideas are given shape; a landscape retains the physical evidence of these mental activities.

- *Contradictions:* Contradictions emerge between human groups because people occupying particular localities develop models of their environment. based on their specific needs and experiences: these models may be at variance with other group, leading to competition, religious conflicts, etc. Contradictions constitute the raw materials for change. Landscapes manifest the resolutions of these contradictions.
- *Scale:* Landscapes cannot be studied in their totalities. Investigation are done at different scales. When a particular scale is chosen during one moment of the analyses, it is because at that effective scale one can comprehend patterns.

Factors of duration, intensity, and periodicity at specific temporal and spatial scales must be examined. These parameters must be considered relative to a specific environment. , because a small change in one environment. could be a major change in another.

- *Culture:* Culture determines the thresholds at which a response will be generated and also the particular response itself. It frames a society's resilience in the face of environmental and other disturbances.

- *Region:* Any area may be termed region for the purpose of the study of human-environment. relationships, so long as demonstrable homogeneity can be recognized. Its temporal relations (connections with the past and the future) and its spatial relations (connections with other areas at the same scale and larger and smaller areas) need to be specified. A region can be recognized at a certain scale because of its distinctiveness from an interrelations with other such units. This pattern simply begins the analysis.
- *Boundaries:* Boundaries have inherent duality. Arbitrary boundaries are to be avoided, except in the initial stage of investigation. Instead a research area defined at a number of different scales, from observation of a multiplicity of not necessarily coincident boundaries, offers a fertile ground fir discovering the contradictions those divisions manifest. Boundaries are interesting both for social and natural scientists to explore relationships with living organisms. There is a high correlation between the position of biomes (and their ecotones) and the position of climate driven air-mass activity. Administrative boundaries do not nest with but overlap environmental, social, and economic boundaries.
- *Biodiversity:* Biodiversity is critical to the maintenance of ecosystems. Richness and innovative potential of cultural diversity can also be seen as a potent tool in the mitigation of human/environment. relationships.
- *Heterachy:* Heterarchical complex systems are systems in which the elements have the potential of being unranked (relative to other elements) or ranked in a number of ways, depending on systematic requirements. Hierarchical systems are systems in which some elements on the basis of certain factors, are subordinate to others. A common error of researchers is to uncritically nest levels of analysis, confusing scalar (global-regional-local) with control hierarchies (court system), and leading to the

misinterpretation of chains of causation. It is quite possible that effects that occur at the "subordinate level" have major systemic effects: they can change parameters (boundaries) or control levels (center-periphery shifts, scale changes) or the ranking of various elements. Heterarchy reminds us of a natural, multidimensional fluctuation in the importance of elements. This flexibility is essential to a dynamic approach.

INTRODUCTORY STATEMENT

We wish to trace in the barest outlines the movement in sociology which has led up to the ecological approach to sociological study. Sociology as a science was conceived when Vico hit upon the idea of producing general social principles and laws by studying and generalizing the data of history and literature. But neither Vico nor the important group of philosophers of history who succeeded him really created an inductive science of sociology. They were still under the spell of the metaphysical concept of natural law, which placed the determination of events in the past and in the distant bowels of the universe, or perhaps in the matrix-mind of divinity. It was Comte who played midwife to a theory of inductive sociology, by insisting upon the determination of events by the interaction and interrelation of the events themselves, instead of by factors or forces (natural law) outside of the events. Thus he stated the formula for all inductive science and named it positivism.

The inductive method had already begun to be used in the older sciences, although it had not been labeled and set in opposition to the older methods before Comte's keen penetration of the significance of intellectual history enabled him to draw the strong contrast between the positivistic and metaphysical approaches. The method of Vico, in the hands of the philosophers of history, had failed to make good its promise of providing social laws and principles. Historical data were not adequate to such generalization. Montesquieu had sought to remedy this defect by the introduction of geographic and climatic factors which he undertook with moderate success to generalize inductively. Buckle combined the methods of anthropogeography and the philosophy of history and elicited a great, if not a permanent, response.

But the death knell of a prioristic social thinking had already been sounded, and, interestingly enough, by the historians. They began to confine their attention to the collection, verification, and storing of facts, largely to the exclusion of their interpretation. All of the other social disciplines followed suit and began to place their main emphasis upon description and the historical method, which was for them a method of fact collection.

But they found it difficult as interpreters of contemporary life to avoid some sort of generalization. This was especially true of sociology. It was never possible after the downfall of the philosophy of history to make of sociology a mere collection of historical data. From its dilemma it found two means of partial escape. One of these was generalization by analogy-especially by biological analogy-in which morass it floundered for nearly a generation.

The other escape was more logical and broke less definitely with the abstemious method of history. Sociology turned in the last third of the nineteenth century to ethnology for the materials it sought to use inductively.

But this recourse was not satisfactory, partly because the results were not much more dependable than those provided by the philosophy of history, and partly because there was an insistent demand that sociology should provide an explanation of life as it is. Already in the eighties the theological schools began to introduce courses in social problems, and this content was merged early in the nineties with sociology.

Along with the content often went the clerical teachers themselves. Their neo-Calvinistic and English-ethical traditions led them rigorously to exclude materialistic factors from their accounts of causes, and even the economic factors they looked upon with suspicion and annoyance. For nearly two decades practically the only mention of physical and economic environment as factors permitted to enter into any serious discussion of social maladjustment came through the scarcely respectable writings of the Italian criminologists.

For the most part sociological theory was almost equally unfriendly to taking cognizance of material environmental factors.

Recognition of the physical environment was relegated to geography, and the recognition of economic factors was either dismissed as socialistic or materialistic interpretation or gladly abandoned to economics, which itself shied at too close contact with this field.

With the exception of a few men like Sumner, the treatment of social processes was in terms of a neo-Hegelian or ethicohistorical ideology. The philosophers and theologues in our camp were numerous.

The rise of the French school of neo-Hegelian psychosociologists in the nineties tended to confirm our own indigenous ideological trend in sociology. Sociology itself came frequently to be defined in terms of psychic interaction among persons, the whole stage on which the action took place being left out of account-at least in theory.

Of course, such an unrealistic sociology could not persist and it fell an easy victim, in its search to escape from a vacuum, to the instinct or biological deterministic interpretation, just as more than a generation before an earlier group of social thinkers, forbidden to generalize directly from insufficient data, had narcotized their imaginations with the biological analogy.

But the seed of a realistic or environmentalist interpretation of human nature and social processes were never lost during these years of the Babylonish captivity of sociology by the shades of Calvin and Hegel. Thomas, at Chicago, in his wide range of reading and interests, brought his students in contact with McGee and the environmental anthropologists.

Sumner worked ahead with his subsistence theory of social origins. And above all, the historians, who in the first half of the nineteenth century, had led the way in the search for facts (which, however, carried the sociologists only as far as uncertain ethnological data), now began at the end of the century to return to nature. Turner and his school are as surely among the forerunners of the new sociology (I mean no offense to them) as are Comte and all the Germans Small lists in his most remarkable Origins of Sociology!

Sociology is today in the process of returning from a number of blind alleys, and it is still a bit dizzy. It is mostly back from its ideological and instinctivist excursions. But the spirit of Hegel still lives, especially through the social mirages of Durkheim and the cultural determinists.

In this last school the surviving influence of Hegel is strongest. A theory of culture which begins in culture and ends in culture, which knows no geography and can bear no environment except that inherited from the past or from some other place-and smothers even this under the term culture-is the latest refuge of the thwarted neo-Hegelian ideologists.

The reason for this trend is, I think, easily found. Such a detached concept of culture is but another name for tradition. Thus the culture interpretation now so popular among sociologists is largely a new disguise for the old ideology and the worship of tradition.

Close aid to the "culture in a vacuum school" is the stratigraphic conception of science, which would lay down the subject matters and methodologies of the various sciences in layers, one above the other, somewhat after the manner of Comte's famous classification. Advocates of this viewpoint apparently would not allow any of the technique or content of one science to trickle through the cement and tile partitions to another.

According to this school of thought a good sociologist should blush to recognize neurons, endocrines, selection (natural or otherwise), climate, or topography and contour. It is not strange, perhaps, that some of our more recently and less metaphysically trained sociologists should take a staff from the hands of the historians and venture to walk on profane ground.

Perhaps it is also not strange that the revolt should have taken an organized form most conspicuously among some of the doctors of just that department where the shade of Hegel has wandered most persistently and where the stratigraphic concept is perhaps strongest.

The new emphasis in sociology upon a realistic approach to its data and methodology is decidedly inductive and environmentalist in contrast to ideological.

The term human ecology is largely accidental and is indicative of (1) the dynamic character of its environmental realism, and (2) of the avenue of its escape from the old vacuum sociology of the ideologists, by analogy with the more dynamic work of the plant ecologists.

The name may or may not last; it may be fortunate or unfortunate. But the realistic approach itself is to the study of the contemporaneous facts and factors which are actually shaping and conditioning human social behavior in an everchanging environment, material and psychic, physical and social. The papers that follow represent a few of the trend along which this new dynamic realism is working out an analysis of the interrelations of man and his environmental controls.

If no papers in the conventional field of population analysis appear on the program of this division this year it is not because no such studies are being made. Perhaps no other phase of sociological research is now so well financed. But most of these studies are strikingly static.

They have not caught fully the spirit of the dynamic trend which lays emphasis upon the analysis of adjustment and changes of adjustment occurring along with changes in social equilibration. We shall not attempt to explain why the population studies have so seldom caught the dynamic impulse to study human and group adjustments which is so characteristic of the trend which calls itself human ecology.

It may be because those who have planned or approved the studies have not caught the dynamic trend. Or it may be something else. But it is a satisfaction to observe in the new movement toward dynamic realism in sociology a promise of the fulfilment of Vico's conception and of Comte's formula for a science of society which should study actual facts of life inductively and from this study discover the sccial habits and organizations of men.

THE NORTH ATLANTIC PROJECT

This proposal seeks to develop the great potential of the temperate North Atlantic Ocean as a system for studying historical ecology, which is the study of ecological communities from a phylogenetic perspective. The scale of this project, intellectually, taxonomically and geographically, requires the large-scale collaborative effort which is the thesis of this proposal. The trans-Arctic exchange of marine organisms in the late Pliocene established many species of Pacific origin in the North Atlantic Ocean. This natural experiment-along with episodes of trans-Atlantic migration of endemic Atlantic taxa-placed many of the same species on both coasts of the North Atlantic.

These coasts differ dramatically in almost every way, including glacial activity, productivity levels, predominant substratum type, and biodiversity.

By comparing the ecology and evolutionary history of sister taxa found on both coasts of the North Atlantic, we can uncover the importance of historical and evolutionary differences which otherwise are obscured by taxon based differences.

In addition, scientists on both sides of the North Atlantic are working to determine anthropogenic effects, such as human-mediated species introductions. Coordination of their efforts would strongly enhance the usefulness of their results for policy makers.

Developing the North Atlantic as a model system for the study of historical ecology has been hampered by two major roadblocks to collaboration. The first is geographical, with American and European scientists only rarely collaborating to study taxa found on both coasts, or to compare the ecology of the northwest and northeastern Atlantic.

The second is the culture of science, where scientists from different disciplines fail to attend meetings in disparate fields, and have difficulty attaining an entree into a literature full of unfamiliar jargon. Our tendency to work independently makes synthesis that much more difficult, since questions of pressing interest to a phylogeographer may not be central to the research program of a paleontologist specializing in the same region.

This proposal seeks to establish an annual meeting bringing together marine ecologists, population geneticists and phylogeographers, oceanographers, paleontologists and paleoclimatologists. These meetings are intended to accomplish five specific objectives:

- o Bring together evolutionary biologists (phylogeographers, population geneticists, and systematists) specializing in marine fish, invertebrates, and algae to coordinate sampling efforts and analysis to synthesize patterns of geographic history held in common among these taxa;
- o Synthesize knowledge about extant diversity on both sides of the Atlantic, exploring productivity and other factors as explanations for trans-Atlantic differences;
- o Encourage collaboration resulting in a series of experiments comparing the ecology of sister taxa, and the parallel ecosystems they populate,on both sides of the North Atlantic;
- o Bring ecologists and evolutionary biologists into contact with the oceanographers and paleontologists whose work is essential to understanding the long-term histories of the communities and organisms involved.

2

HISTORICAL ECOLOGY OF THE NORTH ATLANTIC

HISTORICAL ECOLOGY

Evolutionary biologists define historical ecology as an attempt to understand ecological interactions from a phylogenetic perspective (Brooks 1985; Brooks and McLennan 1991). If two species experience a strong interaction, is the strength of this interaction due to a long shared history, or have they recently come into contact? Although this question has long been a part of ecological debate, phylogenetics has the potential to actually illuminate the "ghost of competition past". For example, consider two interacting species that are suspected of showing niche differentiation, such that one species is large and one is small. If a phylogenetic analysis shows that each species is the same size as its closest relatives from other areas, then it is not necessary to invoke niche differentiation to explain the differences in size.

With respect to community ecology, a phylogenetic framework can help to understand dynamics of community assembly (Cunningham and Collins 1998; Grosberg 2000; Ricklefs and Schluter 1993). Which species have been in residence for a long time (and how long)? Which species have arrived in the recent past (and how long ago)? Are there any generalities about which species have recently arrived?

A Grand Natural Experiment: the Impact of the Trans-Arctic Interchange on the Recent History of the North Atlantic

The temperate North Atlantic is an ideal model system in which to address issues of historical ecology because of a natural experiment in biogeography (Cunningham and Collins 1998; Vermeij 1991). A large proportion of species-approximately 80% of rocky intertidal benthic mollusks in New England-have arrived in the North Atlantic from the Pacific since the opening of the Bering Strait 3-5 Ma. (Marincovich and Gladenkov 1999) (Vermeij 1991). This trans-Arctic interchange united the cold water biotas of the Pacific and North Atlantic/Arctic basins, which had evolved independently until that time.

Like most other biotic interchanges, the direction of interchange was highly asymmetric (Pacific to Atlantic, Figure 1(Vermeij 1991)). This natural experiment placed closely related species throughout the northern hemisphere, on coasts which differ dramatically in almost every way.

This experiment allows direct comparison of the ecology of these closely related taxa, so that observed differences between regions cannot be attributed to taxon based differences. For example, the crustacean isopod Idotea balthica is found in a broader range of algal and seagrass habitats, in New England populations than in European populations, and plays a more dominant role in the New England community.

The differences between the northwestern and northeastern Atlantic can be used to investigate community assembly. The last glaciation was much more extreme in the NW Atlantic than in the NE Atlantic, and is thought to have caused local extinction of many benthic species (Vermeij 1989). In some cases, these extinctions may have been followed by post-glacial colonization from the NE Atlantic (Cunningham and Collins 1998). Population genetics, in a geographic context (phylogeography, Avise 2000), has the potential to distinguish between taxa that did or did not survive the last glaciation in the NW Atlantic (Wares and Cunningham in revision). Once the recent colonists have been identified, we can ask which are likely to have been natural or human-mediated introductions, and search for generalities such as what organismal characteristics allow species to survive glaciation in the NW Atlantic. Despite this incredible potential, there have been virtually no comparisons of species found in both coasts of the North Atlantic. Rectifying this situation is one of the central purposes of the North Atlantic Project.

Three Views of Generality in Historical Ecology and Biogeography

A major outstanding question in historical ecology, and biogeography in general, is what role organismal characteristics play in determining composition of communities. Although it may seem obvious that organismal characteristics determine range limits of species, the issue of how those species came to be found in a particular region is open to question.

Interestingly, two completely independent intellectual traditions in Biogeography downplay the importance of organismal characteristics. The first is island biogeography (MacArthur and Wilson 1967), and its intellectual descendants (Hubbell, 1997,2001). This tradition considers raw species numbers, and considers individual species largely interchangeable. Although dispersal and extinction are both central to these theories, the difference in dispersal abilities between individual species are assumed to average out in the end.

The second great intellectual tradition is vicariance biogeography (Humphries and Parenti 1986; Nelson and Platnick 1981). Whereas dispersal is central to island biogeography, in vicariance biogeography the emphasis is on the signal of past fragmentation that can be gleaned from the species that do not disperse from their areas of endemism. If species collected from the same areas show congruent phylogenies, this is considered evidence that they experienced the same series of vicariance events (geographic subdivision). Vicariance biogeography emphasizes huge events that affect entire biotas, regardless of individual organismal characteristics. In this sense, it is similar to island biogeography in that it seeks generality by downplaying organismal characteristics.

In contrast, a third view considers the possibility that organismal characteristics are important. Diamond's controversial assembly rules (1975) argued that the composition of many communities depends on organismal characteristics (see also the forthcoming book "Ecological Assembly Rules" (Weiher and Keddy 2001)). In this vein, Cunningham and Collins (1998) have argued that phylogenetic history has the potential to reveal the influence of organismal characteristics in biogeographic histories. If we are able to identify a set of species that have experienced a common biogeographic history, we can ask whether they share particular organismal characteristics. While this may sound at first like vicariance biogeography, it doesn't assume that there is a single overriding historical pattern (determined by vicariance) that defines an entire biota. Common patterns of dispersal, such as the trans-Arctic interchange itself, can generate patterns as well. In marine systems, with their porous barriers, it is unlikely in most cases that vicariance will affect more than a subset of the biota (with obvious exceptions such as the rise of the Isthmus of Panama). In many cases, we expect that subsequent dispersal between vicariated areas will be extremely common (Cunningham and Collins 1994).

With respect to the North Atlantic, this perspective can be framed in terms of two likely patterns. The first is glacially induced extinction in the NW Atlantic. If a species experiences local extinction without recolonization from elsewhere, the range of this species has been restricted. Vermeij(1989) pointed out that paleontology is the only discipline that can unambiguously identify cases of restriction. Cunningham and Collins (1998) argued that population genetic and phylogenetic data can accomplish the converse, and unambiguously identify sets of species that have long endemic histories in the NW Atlantic, and have therefore resisted glacial extinction (Cunningham and Collins 1998 Wares and Cunningham, in revision).

Since repeated extinctions and recolonizations may appear in a low resolution fossil record as a coninuous presence, genetic data are important in this context. In high resolution mcrofossil records, these repetitive local extinctions and recolonizations can be directly observed. These stratigraphic records can then be compared to the conclusions of phylogeographic studies. The first observed pattern is one of long term residence on both coasts of the North Atlantic. Homarus gammarus in the NE Atlantic and Homarus americanus in the NW Atlantic are sister species, and have had independent evolutionary histories for roughly 1 million years or so (Tam and Kornfield 1998). This "reciprocal monophyly" is strong evidence that Homarus has been in residence in both the NE and NW Atlantic for at least 1 million years (Cunningham and Collins 1998). What is it about Homarus that has allowed it to resist extinction in the NW Atlantic through several glacial maxima? The second is a population genetic pattern shown by many species strongly suggesting post-glacial colonization of the NW Atlantic region from the NE Atlantic (e.g. the seastar Asterias rubens, and the snails Littorina obtusata, and Nucella lapillus (Wares and Cunningham in revision)). What characteristics allowed these species to invade across the Atlantic? Since the latter two species have no pelagic dispersal, dispersal ability is not the only important factor in colonization ability. Other factors inlude the ability to withstand the wide temperature fluctuation in the NW Atlantic, and the tendency to occupy dispersal vectors such as floating algae (Ingólfsson, 1995).

SPECIFIC AREAS OF RESEARCH COORDINATION

Although the range of fields represented in this RCN are very broad, we envision specific areas of concrete, mutually advantageous coordination. Through our regular meetings and through classes of information that will be collected at our website, we envision a rapid improvement in our knowledge of North Atlantic biology in the next five years.

Coordinated Phylogeography and Systematics of the North Atlantic Biota

By documenting the evolutionary histories of the temperate North Atlantic biota-and searching for sets of species with similar biogeographic histories-we can ask whether organismally-based generalities exist. For robust patterns to emerge, we need an ambitious coordination effort that will investigate hundreds of species in a comparable framework. This would be the first time such a comprehensive effort has been attempted for any community on earth. We already have 30 laboratories interested in coordinating our phylogeographic efforts, and we expect this number to grow rapidly as our network expands.

The sheer number of cooperating laboratories is not as important as the taxonomic breadth. For sociological reasons, there is little communication between phylogeographers working with different taxonomic groups. For example, although algal biogeographers have an impressive history of research coordination and synthesis (consider the pioneering volume Evolutionary

Biogeography of the Marine Algae of the North Atlantic (Garbary and South 1989), and BIOGAP (biodiversity and genetics of algal populations, coordinated by network member M. Valero) there is virtually no communication between phycological phylogeographers and zoologists-despite the fact that our organisms undoubtedly share many of the same sets of biogeographic histories.

Even more than between plants and animals in terrestrial systems, there is an enormous commonality between the experience of benthic invertebrates (particularly when they are sessile) and marine algae, especially with respect to modes of dispersal. Even among zoologists,communication between fish and invertebrate phylogeographers and population geneticists needs improvement. Finally, foraminiferal specialists have historically been lodged in geological departments and only recently have started to develop cooperation with biological and genetic researchers.

Towards a Comprehensive Phylogeography of the North Atlantic Biota Our coordination in this regard has four major goals:

- *Eliminate Redundancy:* Assembling this network has already revealed that three independent groups have been working on trans-Atlantic genetic studies of the gastropod Nucella lapillus!
- *Inspire Larger Scale of Analysis:* We need to think much larger than the few species we are currently studying. With routine PCR and sequencing, multiple species comparisons will become routine. By distributing responsibility for large parts of the trans-Atlantic biota, large-scale projects will become a reality.
- *Use the Same Molecular Markers:* Where possible, we should use the same markers to allow combination of independently collected data sets.
- *Increase efficiency of Broad Geographic Sampling:* We can coordinate our collecting efforts in two ways:
    - o First, individuals at cooperating marine labs can collect "wish lists" of species from individuals in the network. Where possible, modest compensation by the requesting laboratory will be provided to the collectors.

This will allow a more uniform sampling regime across taxa, and allows those individuals most familiar with the local biota to do the actual collecting, which increases efficiency, and increases the likelihood of finding rare species.

- o Second, when collecting trips are planned-particularly those involving research vessels-collectors will canvas members of the network via email and through our website, for wish lists of taxa.

Although phylogeographic studies are generally focused on variation within species, they usually turn up cryptic species not suspected by morphologists. Our phylogeographic efforts will be carefully integrated with the efforts of our strong team of systematists and experts in biodiversity as described in the next section.

Re-evaluating the speciational history and taxonomy of the North Atlantic To fully understand the evolutionary history of the North Atlantic, we must put studies of variation within species in the context of the origin of new species. This is impossible without re-evaluating the alpha taxonomy of North Atlantic species. As described below, the discovery of cryptic species in phylogeographic studies is very important in this regard. Our network includes individuals who have already spearheaded large coordinated efforts in this regard, and we hope to inspire other workers to emulate their work.

The Virtual Handbook of Trans-Atlantic Species

Because existing handbooks are restricted to either the American or European coasts, it is difficult to determine which taxa are trans-Atlantic. Even when a species with the same name is found on both coasts, handbooks rarely mention the full extent of the species range, so that a collector won't be immediately aware that a species they have found is also found across the Atlantic. Compiling a list of trans-Atlantic species is complicated by two issues: synonymy and cryptic speciation.

- *Synonymy:* For historical reasons, the same species is sometimes given different names on each coast. For example, the charismatic mega-invertebrate seastar genus

Asterias had three named species in the North Atlantic; A. forbesi and A. vulgaris in North America, and A. rubens in Europe. A morphological analysis found that A. vulgaris and A rubens were synonymous, and proposed that American A. rubens populations are recent colonists from Europewere recently founded from Europe (Worley and Franz 1983). This has been confirmed by molecular analysis (Wares in press a; Wares and Cunningham in revision).

- *Cryptic Speciation:* In many cases, morphologically indistinguishable taxa are actually deeply diverged, and according to many species concepts, should be elevated to species status (Knowlton 1993; Knowlton and Weigt 1995). For example, in the North Atlantic, the isopod Idotea balthica is actually composed of at least three deeply diverged lineages, one of which is trans-Atlantic (Wares in press b; Wares and Cunningham in revision).

Inspiring and Coordinating Trans-Atlantic Ecology Studies

If the North Atlantic is to realize its potential as a model system for historical ecology, we must take advantage of the replication proved by hundreds of trans-Atlantic species. The advantage of studying shared taxa under the different ecological settings of the NW and NE Atlantic is that it provides insight into how differences in community structure influence the distribution and abundance of species, while also allowing for the possibility of local adaptation. It also provides insight into the consequences of community change brought about by species introductions or extinction. Indeed, one of the most exciting prospects for comparative trans-Atlantic ecology would exploit the existence of particular ecosystems (e.g., eelgrass beds, rockweed beds, mudflats) common on both sides of the Atlantic which are very similar apart from the NW Atlantic biota being a consistently lower diversity subset of the EaNEstern Atlantic biota.

This trans-Atlantic contrast in biodiversity provides a natural experiment, on a grand spatial and temporal scale, for addressing the pressing issue of how biodiversity loss is likely to influence the structure and function of coastal marine ecosystems. Such comparisons have rarely been attempted (but see the next section), in large part because they will require just the sort of large-scale, coordinated effort proposed here. We will model our cooperation on the EUROROCK programme which did the same experiments on grazer/algal interactions and barnacle recruitment throughout Europe, which was coordinated in part by network member Stephen Hawkins.

Preliminary trans-Atlantic ecological comparison of decapod crustaceans:

The shallow marine fauna of the European Boreal is richer than the American Atlantic Boreal. For example, there are 41 brachyuran crab species in this part of Europe (Christiansen 1969) versus only 10 in the Gulf of Maine and Canadian Maritimes (Williams 1984). Similarly, in the North Sea there are 18 gadid and more than 6 labrid fish species, but only 12 and 2, respectively in the Gulf of Maine (Williams 1984).

Steering committee member R. Wahle found that in the Gulf of Maine, H. americanus and Cancer irroratus, the rock crab, were by far the two most abundant subtidal species in cobble, with C. borealis, the Jonah crab, typically less abundant. The introduced green crab, Carcinus maenas, was more abundant intertidally. By contrast, in Ireland and the Channel Islands, Homarus and Cancer were the least abundant genera of the European decapods. Four other decapod families dominated: xanthid crabs, galatheids, porcellanids, and alpheids with at least two to three species of each.

Very little otherwork has been done directly comparing the ecology of trans-Atlantic marine organisms. The only published work that we know of is by network members Lotze and Worm (2001), who compared the ecology of Baltic and Canadian macro-algae. Inspiring trans-Atlantic Ecological Studies

Although the International Council for the Exploration of the Sea (ICES) regularly brings together applied ecologists from both sides of the Atlantic, there is no mechanism to bring together basic ecologists in the same manner. We envision three ways to encourage trans-Atlantic ecological research

- o First, our virtual handbook of trans-Atlantic species (see above), will draw attention to species shared between the two coasts of the Atlantic.

- o Second, each year network members will visit the American Marine Benthic Ecology Meetings and the European Marine Biology Symposium to recruit ecologists to carry out cooperative trans-Atlantic studies. As preliminary results accumulate, symposia will be organized at both of these meetings to highlight trans-Atlantic ecological research.
- o We will post an electronic "bulletin board" of ongoing research projects on either side of the North Atlantic, to identify researchers with interests and preliminary research on trans-Atlantic species.

Reciprocal Illumination between Ecologists, Phylogeographers, Systematists and Paleontologists

Coordination is useful not only within but between disciplines. When ecologists and phylogeographers are studying the same species, they can meet, and teach one another aspects of the species' biology and history that will deepen their understanding of their organisms. If an American ecologist learns from a phylogeographer that a particular species has colonized America in the past 20,000 years, that researcher can investigate the possibility of a bottleneck affecting the ecology of that species. Conversely, if an American ecologists learns from a systematist that a trans-Atlantic species is actually composed of two deeply diverged cryptic species, then the possibility of local adaptation on either side of the North Atlantic must be investigated. Fossil records of taxa can establish were species first were observed, thus possible the location of origin of the modern taxa.

Phylogeographers will be encouraged to study the set of taxa that will be targeted for trans-Atlantic ecological experiments. The annual meeting and electronic bulletin boards will help to bring these researchers together.

Educating Phylogeographers and Systematists about Geology and Oceanography

Although the evolutionary history of marine species cannot be understood without knowledge of geology and oceanography, few phylogeographers and systematists are trained in either discipline. As a result, the literature of these crucial fields can be opaque to those from other disciplines. This problem is compounded by the fact that questions that are of interest to phylogeographers and systematists may not be part of the research agenda of geologists and oceanographers. Each annual meeting of the North Atlantic Project will feature workshops to bring together scientists from these disciplines. This will also give geologists and oceanographers the opportunity to learn about, and to give possible explanations for, patterns found by the phylogeographers and systematists.

For example, the observation that many species of the NW Atlantic have colonized from the NE Atlantic since the last glacial maximum must be reconciled with information from paleoclimatology and paleo-oceanography (Cunningham and Collins 1998; Wares and Cunningham in revision). Similarly, the observation that some species with long fossil histories in the NW Atlantic may have experienced local extinction during glacial maxima followed by recolonization from the NE Atlantic (e.g. Nucella lapillus and Littorina obtusata:Wares and Cunningham, in revision) may inspire paleontologists to ask whether this lack of continuity is evident upon a closer examination of the fossil record. Also, as discussed above, paleontologists are uniquely suited to identifying cases of geographical extinction (Vermeij 1989).

The workshops at our annual meeting will give an opportunity for workers from these diverse disciplines to directly question one another, and to develop relationships that will allow them to delve intelligently into an unfamiliar literature.

ANNUAL NETWORK MEETINGS IN AMERICA AND EUROPE

The North Atlantic Project will meet annually, alternating between locations in America and in Europe.

First meeting is scheduled for August 20th-23rd 2002 at Shoals Marine Laboratory in New Hampshire, hosted by network member and director of the Shoals Marine Laboratory, J. Morin;

Second meeting is scheduled in Reykjavik, Iceland, hosted by steering committee member A. Ingólfsson. Given that Iceland is literally central to coordinating North Atlantic research, this will help increase involvement of Icelandic scientists in our effort. We envision these first meetings as being fairly small (60-80 participants), to allow easy communication between disciplines, and to prevent concurrent sessions from fragmenting our group. The question of how large to allow the meetings to become is of great importance, and will be decided by steering committee. To reduce fragmentation of the group, most presentation of scientific results will take place during poster sessions, which in a small group will work to cement individual relationships.

These meetings will include workshops approved by the steering committee, but topics will include:

- o Coordinating trans-Atlantic phylogeography and sample collection;
- o How to identify introduced species from phylogeographic data;
- o Coordinating trans-Atlantic ecological projects;
- o Constructing electronic bulletin boards to encourage trans-Atlantic and crossdisciplinary collaboration;
- o Paleo-Oceanography of the North Atlantic;
- o Constructing the virtual handbook for trans-Atlantic species, and integrating existing databases of biodiversity;
- o Coordinating research teams seeking joint funding from the United States, Canada and the European Union.

THE NORTH ATLANTIC PROJECT STEERING COMMITTEE

Twenty scientists from eight countries bordering on the North Atlantic have agreed to serve on the steering committee, including representatives of every major discipline included in our network (see project summary for list). This group does not include all researchers who have indicated their enthusiasm in playing a central role in our coordination network. We see the role of the Steering Committee as being primarily administrative, as we will regularly seek input from network members on all topics of scientific importance such as designing workshops and facilitating research. Nevertheless, a project of this scale must be carefully organized by a group of manageable size. The main functions of the coordinating committee will be:

- o To meet at Duke University in the Spring of 2002 to plan the first meeting at Shoals marine laboratory. Subsequent meetings of the Steering Committee will take place as part of the annual meetings, and will include preliminary plans for the following meeting;

- o To approve topics of workshops suggested by network members;
- o To approve applications for travel funding from network members, and to manage the funds sought in this proposal. At first, this will be restricted to funding travel for American scientists and meeting costs for all attending scientists. Additional travel funding will be sought from Canada and the European Union for international collaborators.
- o To decide the locations for the next three meetings. This decision will include the maximum size of each meeting.

Size of Annual Meetings and Expanding the Membership of the North Atlantic Project

The North Atlantic Project cannot function without a large number of participating scientists. Our commitment to inclusivity is illustrated by the large size of our steering committee, who are charged with expanding the membership of the North Atlantic Project. On the other hand, larger meetings will inevitably become fragmented according to discipline, which is explicitly contrary to our central purpose. We will attempt to resolve these conflicting forces in three ways that highlight inclusivity.

- o The annual meeting is only one way in which scientists can contribute to the North Atlantic Project. Our website and listservs will include summaries of ongoing research and invitations to cross-disciplinary and trans-Atlantic collaboration;
- o Recruiting interested scientists by organizing symposia at regularly scheduled meetings, including (in North America) the Evolution meetings, the Ecological Society of America Meetings, the Marine Benthic Ecology Meetings, and the meeting of the Society of Integrative and Comparative Biology; (in Europe) the annual meeting of the European Evolutionary Society, the October 2002 meeting of the Italian Biogeography Association in Ischia, Italy, and the annual meeting of the European Marine Biology

Symposium. To improve communication with geologists, representatives of the network will make presentations at the foraminiferal FARMS 2002 meeting, and at the North American Paleontological Convention, and the International Paleoceanographical Meeting.

- o If the size of the annual meeting is not enough to include all interested participants, we will give preference for travel funding to scientists who have not yet attended an annual meeting.

Recruiting a Young and Diverse Coordination Network

Recruiting a new generation of workers to trans-Atlantic research will be an easy task. Many of the scientists in our network with the most direct experience in trans-Atlantic research are at the post-Doctoral level (e.g. T. Dahlgren; G. Trussell; B. Worm; H. Lotze; represented on steering committee by post-doc J. Wares). The promise of exciting large-scale research questions should attract a large number of young workers.

The network has a strong commitment to increasing participation from under-represented groups. The leadership role played by women in North Atlantic research is reflected in the high proportion of women in our steering committee (38% vs. 20% in our network as a whole). PI Cunningham has a demonstrated commitment to encouraging minority representation. He currently has Hispanic-American and African American undergraduates working on North Atlantic research in his laboratory; the latter, Audrey Abrams, attended the 2001 SSE meetings in Knoxville supported by the Diversity Committee of the SSE, and is seriously considering a career in evolutionary biology.

We also envisage communication of our results and use of our web site in pre-college education. Steering Committee member Thomas has worked in adult education and is active in developing teaching material in direct cooperation with middle and high school teachers, within the Program for Increase of Mastery in Mathematics and Sciences (PIMMS) at Wesleyan University. Steering committee member Bucklin is working with to develop a high school curriculum module for molecular systematics aand marine biodiversity for her Zoo Gene website.

To encourage young scientists, and scientists from underrepresented groups, PI's will be encouraged to bring at least one colleague to the meetings who belong to one of these categories. In allocating travel funding, the Steering Committee will favor their applications.

FUNDING ALLOCATION

Allocation of Funding to American and International Collaborators

The RCN program intends for international collaborators to seek their own funding to attend meetings. Therefore, this grant would primarily fund travel by American scientists, and fund accomodations and meeting costs for all scientists.

Michael Hart and Mark Costello are planning to seek Canadian funding, and several network members have expressed a desire to apply for EU funding.

In the interim, most international members have indicated that they have enough funds to travel to these meetings. Because it is unreasonable to ask international collaborators to travel twice to the US during the first year, we will ask for funds to help steering committee members travel to the organizational meeting at Duke University.

Administrative Personnel at Duke University and Supervision of Website

The North Atlantic Project will be coordinated by Cliff Cunningham at Duke University, who will supervise a Duke graduate student responsible for maintaining the website and for handling correspondence and travel reimbursements in coordination with Duke Biology accounting staff.

RELATION TO EXISTING EFFORTS AT RESEARCH COORDINATION

Although this proposal has emphasized the coordination efforts we will initiate, we are well aware of, and inspired by, the many ongoing efforts at international and cross-disciplinary coordination in North Atlantic research. Many of these efforts are aided or coordinated by members of our own network.

Applied Ecology

For applied ecological problems, the International Council for the Exploration of the Sea (ICES) is multi-disciplinary-bringing together oceanographers and fisheries biologists-and is explicitly trans-Atlantic in its scope.

Its efforts overlap to some extent with the aims of this proposal (both are interested in identifying human-mediated species introductions), but in general do not extend to basic ecology or evolutionary studies.

3

Historical Ecology and Future Explorations

INTRODUCTION

There are no environmental limitations to the development of agriculture, only cultural limitations. "Agricultural potential" is a cultural phenomenon; it is not something inherent in nature that can be measured, that exists independent of culture. Today, with available technology, agriculture can be carried out anywhere on earth. Whether it is or not in any given habitat is dependent on whether the culture involved has the necessary technology, and whether or not there is a perceived need in relation to the costs involved (capital, material, labor) (Denevan, 1982: 181).

Soils in Amazônia are traditionally thought of as a given, limited homogeneous resource labeled as poor or marginal for agriculture. Historically and ethnographically, the strategy of settlement and farming in Amazônia is to colonize new farmland (a form of agricultural expansion) and use them for 2-3 years until crop yields decline and/or weeds become too much of a problem, abandon the field and move on to a new location (e.g. slash-and-burn agriculture and now increasingly modern ranching in the tropical forest). Agricultural intensification, increasing yields through addition of labor, fertilizers, irrigation (infrastructure), mechanization, and capital to unit area of production has had limited success in Amazônia.

Amazonian Dark Earth (ADE) may represent a significant case of agricultural intensification by Native Americans, the artificial creation of new, highly productive soils. The investment of energy, materials, and resources into permanent farmland near settlements apparently had positive results based on the archaeological evidence for size and longevity of occupation on these sites. The form and internal variation of ADE provides insights about patterned human 455 J. Lehmann, et al. (eds.), Amazonian Dark Earths: Origin, activities within pre-Columbian settlements that produced them. Research on ADE may also contribute to contemporary rural development.

Historical Ecology provides an interesting framework to discuss Amazonian Dark Earth (ADE). The basic premise is that Native Amazonians did not adapt to nature, but rather they created the world that they wanted through human creativity, technology and engineering, and cultural institutions (Balée, 1998; Denevan, 2001, Stahl, 1996; Erickson, 2000a). Focus is on the human history rather than natural history of the environment contextualized within an historical and cultural tradition. For example, Balée's "model of agricultural regression" (1994) states that present-day native farmers and foragers live off resources in their environment that were created by the efforts of hundreds of generations of previous occupants of the landscape.

Humans do not adapt to a given natural environment, but rather are exploiting the past, complex human history of the landscape; partaking in the accumulated landscape capital created by their ancestors. Rather than passive responders or adaptors to a given environment as stressed in the traditional approaches, historical ecologists view humans as active agents in their interaction with nature who promote change and continuity through culture. The landscape, the central conceptual structure for historical ecology, is the physical manifestation of the long-term human history of the environment. The goal is to document and understand the long-term creation of the environment as we know it today. This understanding may provide models for strategies of conservation and management of the environment in the present and future.

ADE is an excellent example of an anthropogenic (human-created) landscape feature, one of many past human activities permanently embedded in the landscape that continue to structure the environment and are still exploited by native peoples today.

In fact, ADE was prominent in two classic publications that critique the widely accepted idea that Amazonian tropical forests are pristine wilderness (Balée, 1989; Denevan, 1992a).

It is the key role that soils have played in the history of thinking about the past, present and future of the Amazon Region that makes this volume about ADE so timely and important. ADE was recognized by scholars in the last century, but remained relatively unknown to Americanist scholars until the 1980s (Myers et al., 2003; Woods, 2003).

The volume surveys the history of research about ADE and presents detailed analyses and new interpretations about the content, origin, use, maintenance, and sustainability of these anthropogenic soils (anthrosols). The authors of this volume demonstrate that Amazonian peoples were capable of transforming soil horizons, improving nutrient availability and recycling, altering hydrology, changing vegetation succession, and capturing sediments; but they also created or constructed new productive soils. These terms "create" and "construct" capture the scale and significance of what Amazonian people did to improve their environment for human existence and development. Whether these regional-scale anthropogenic impacts were negative or positive over the short and long term continues to be debated and will be discussed in a later section of this chapter.

Soils have long played a key role in the structuring of our understanding of past and present Amazônia. When humans and soils are discussed at the same time, it is usually framed in terms of environmental limitations or possibilities or progressive human degradation of landscapes through overuse and erosion.

In trying to understand why civilizations rarely developed or if present, soon collapsed in regions of tropical forest, scholars (Steward, 1948; Meggers, 1954, 1971; and others) proposed that poor soils were environmental limitations to cultural development. According to the Meggers' "Theory of Environmental Determinism" (1954), swidden agriculture (slash-and-burn agriculture, shifting cultivation) developed as an adaptation to these poor soils (summarized in Myers et al., 2003). Because classic swidden requires large tracts of forest for the periodic establishment of new fields after abandonment of exhausted plots, Amazonian societies could never maintain sufficient population density for the development of complex societies.

The relatively low social complexity, nomadism of foragers and small villages of mobile slash-and-burn farmers of the ethnographic record were recognized as modern day representative analogs of static cultural development. Based on ethnography, Carneiro (1960, 1961) responded that manioc production under slash-and-burn could be quite productive and sustain large, permanent villages of 1000-2000 people. Small group size, frequent settlement change, and endemic warfare was blamed on protein scarcity, added to the list of environmental limitations of Amazônia in the 1970s (e.g. Carneiro, 1970; Gross, 1975). Meggers softened her view in later publications about environmental limitations, but continues to deny archaeological evidence for dense populations and complex societies in the Amazon (e.g. Meggers, 2001).

Few contemporary scholars accept the hypothesis of environmental limitations and lack of cultural development in the Amazon Basin. Archaeological, ethnographic, and historical research over the past 30 years has demonstrated that Amazonian peoples developed complex societies, developed sustainable and often intensive agriculture, packed people on the landscape for thousands of years, and built the environments they desired. The research on ADE provides important additional evidence that shows that Meggers and followers were incorrect. They made the mistake of assuming that peoples recorded in the ethnographic record, in particular their subsistence practices and level of socio-political complexity, were representative of the past.

Meggers and colleagues fall into the trap of assuming that swidden agriculture practiced today throughout the Amazon to be typical of the use of these landscapes in the past. Lathrap et al. (1985: 54) came to the conclusion many years ago that "shifting slash and burn agriculture was a secondary, derived, and late phenomenon with in the Amazon Basin". According to Lathrap (1987), the labor for opening up forest using the slash-and-burn technique became worthwhile only with the introduction of productive and storable races of maize. More recently, Denevan (1992b) argues that slash and burn was probably not common until the introduction of metal axes and machetes by Europeans; thus, the extensive agriculture practiced today on ADE and elsewhere tells us little about original pre-Columbian practices. Authors of this volume conclude that slash-and-burn did not contribute to the formation of ADE.

Instead of slash-and-burn, Lathrap (1970, 1987; Lathrap et al., 1985) pointed out that good agricultural land is found on the annual flood-enriched soils of levees of the river floodplains. Farmers in the seasonally inundated savannas built raised fields (Denevan, 1966, 2001; Erickson, 1995; Walker 2003) (Fig. 1). Carneiro (1960, 1961, 1970, 1996) and Lathrap (1970, 1977) highlighted the importance of manioc and fish in sustaining large sedentary populations. Others highlighted the importance of palms and fruits as important sources of protein in addition to fish (Beckerman, 1979; Clement, 1999a, 1999b). Roosevelt (1980, 1991) showed that the introduction of maize some 2000 years ago created conditions for the rise of complex societies in the central and lower Amazon. The contribution of agroforestry for permanent production throughout the Amazon Basin was recognized in the 1980s (Denevan and Padoch, 1988; Posey and Balée, 1989).

Authors of this volume have presented detailed histories of ADE research (Glaser et al., 2003; Mora, 2003; Myers et al., 2003; Sombroek et al., 2003; Woods, 2003; also Smith, 1980; Mora et al., 1991; Denevan, 2001; and others). The history of research shows an interesting cycling of discovery, neglect, and rediscovery of the importance of ADE since the earliest references in the 19th century. Referring to his now famous article about ADE published in 1980, Nigel Smith states "I got two reprint requests for that article. Nobody was ready to hear it" (N. Smith quoted in Mann, 2000b). Since the 1960s, ADE studies have traditionally been the domain of archaeologists, ethnographers, ethnobotanists, geographers, and soil scientists working individually. Since the late 1980s, most ADE research has been increasingly interdisciplinary and multidisciplinary as highlighted in this volume and recent publications, symposia, conferences, workshops, and popular press (Mann, 2000b, 2002a, 2002b).

The contributors of this volume explicitly or implicitly grappled with the paradox facing all archaeologists who attempt to explain and understand the past. The peoples and processes that produced ADE existed in the distant past while our systematic observations (archaeological and soil stratigraphy and artifact analysis, ethnographic and historical analogy, experiments, and laboratory analyses) of ADE are done in the present, temporally removed from the processes that produced them.

In order to understand the past from present day observations, systematic "bridging arguments" or a "middle range theory" is needed to link the past and the present. In this volume, various forms of explicit analoga are employed (e.g. archaeological, ethnographic, historical, experimental and cross-cultural). Despite the problems inherent in the misuse of the analoga discussed above, I agree with many authors of this volume who argue that we can learn important lessons about the origins and formation of ADE through constructing testable models based on experiments, archaeology, history and ethnography.

Assuming continuity of traditional agriculture from the pre-Columbian to present ignores the historical dimensions of conquest and post conquest policies on native peoples. As authors of this volume point out, Native Amazonia was massively depopulated during the Colonial period and the large and complex societies of the central Amazon Basin disappeared, along with much of their knowledge system, technology, cultivation strategies, and crops (Clement et al., 2003; Myers et al., 2003).

DEFINING AMAZONIAN DARK EARTHS

The authors in this volume are in agreement that the basic characteristics of ADE are their dark color, richness in charcoal-derived carbon, high fertility, and human origin. Beyond this, there is little agreement.

The diversity of approaches for characterizing and identifying ADE are highlighted in this volume. The authors concur that ADE should not be defined too narrowly because of the rich variation within and between ADE (Kämpf et al., 2003). The lack of a single definition or clearly defined suite of characteristics for identifying ADE is obviously frustrating for some of the participants in this volume. At this point in the study, flexible definitions that recognize the variation are healthy.

How is ADE soil identified? The principle criterion is dark color. Various issues can be raised about color. How dark do soils have to be in order to be identified as ADE? Is a simple threshold on a Munsell color chart reading sufficient? Could soils be of radically different colors but have the same composition and formation process? What "natural" soil benchmark is used for comparison? At what point does archaeological soil (soil of archaeological settlements, monuments, middens, and earthworks) become ADE? Are ADE without the presence of archaeological artifacts anthropogenic? The authors of this volume grapple with these questions and provide some answers. In an attempt to address the issue of continuous distribution of soil color from brown or gray to black, scholars working in the central Amazon Region (Sombroek, 1966) introduced the term terra mulata ("brown soils") for the large transitional zone around terra preta (the classic ADE).

The diverse meanings of ADE and potential problems of communicating between scholars of different disciplines were clearly represented in the 2002 TPA Workshop and in this volume (Kämpf et al., 2003). By relying on a vague definition of ADE, we potentially open ourselves to charges of over-or underestimating the geographic extent and importance of ADE. On the other hand, a overly narrow definition might ignore the rich variation of ADE reported in this volume. Kämpf et al. (2003) consider the dynamic, historical and variable nature of ADE as a subset of general anthrosols (soils produced through human activities) for their Archaeo-pedological Classification. Their new classification is an attempt to combine insights from various disciplines to address the variability and continuous variation of ADE.

Archaeologists, who identify and map ADE, tend to rely on the discipline's soil classifications and interpretations that rarely agree with those defined by modern soil science. Archaeologists are experts at recognition and interpretation of a wide range of anthropogenic soils associated with human activities on sites and landscapes. Soil scientists focus on the horizons in the profile and consider anthropogenic features as noise, perturbation and disturbance. In contrast, archaeologists define the visually and/or texturally obvious anthropogenic features in the profile and treat the horizon formation as noise, perturbation, and disturbance.

To the archaeologist, "natural" soils are only interesting in terms of defining the boundaries of the anthropogenic soils (i.e. site boundaries; sterile boundary under site, and so forth). Archaeologists are most likely to focus on the internal variation of ADE: faint patterns, changes of texture, color, context, fill, and features to extract function and meaning. Without the internal heterogeneity within a site, we could not do archaeology. Soil scientists are

less concerned with these soil nuances and attempt to characterize features representative of larger spatial areas. The approach proposed by Kämpf et al. (2003), and to a certain degree, traditional archaeology dedicated to building chronologies, focuses on profile descriptions of small, often deep, excavations through ADE (or sites) which emphases vertical continuity and disjuncture. Since the 1970s, archaeologists have used large areal excavations often combined with sampling to recover spatial patterning of human activities and lifeways throughout the site, emphasizing a horizontal perspective. Kämpf et al. (2003) discuss the contrasting approaches used by archaeologist and soil scientists in regard to ADE and highlight the benefits of combining both approaches. ADE research can benefit from both the general chronological approach and spatial morphology of settlement and agriculture approach (e.g. Heckenberger et al., 1999; Petersen et al., 2001; Kern et al., 2003; Neves et al., 2003; and others). I agree with the Kämpf et al. (2003) plea for collaboration between archaeologists and soil scientists.

The practice of traditional archaeology is framed within the site concept (e.g. critiques by Dunnell, 1992; Fotiadis, 1992). A site is a basic discrete unit of analysis defined by concentrations of artifacts indicating settlement or other activity assumed to reflect human behavior. I argue that adherence to the "site concept" limits our understanding of historical ecology in the Amazon Basin. The concept of landscape within Historical Ecology and archaeology of landscape are powerful alternatives to site-based approaches and are what links archaeology to historical ecology.

Rather than focus on arbitrarily defined sites, landscape approaches try to understand human activities that occur between traditional sites and across larger areas at multiple scales. In this perspective, human activity is viewed as continuous over the landscape rather than spatially contained within traditional sites. Despite new innovative methods for archaeological survey, recovery of human residues, and "non-site" landscape approaches for defining between-site human activities (e.g. Stahl, 1995), Amazonian archaeologists are drawn to "sites," usually pre-Columbian settlements, defined by conspicuous surface concentrations of pottery, lithics, and charcoal (the most commonly preserved archaeological materials).

By extension, ADE research has adopted the site concept. Are ADE discrete spatial units of analysis as presented in this volume? How can a typical black earth site be measured if it has no clearly defined boundaries or edges? Most archaeologists and historical ecologists now recognize that the earth's surface is covered with continuous distribution of artifacts and evidence of human transformation of the landscape, making it difficult, if not impossible to clearly define boundaries. For example in the agroforestry literature on the Amazon Basin, every landscape has been transformed to some degree by thousands of years of human activities (burning, selection for economic species, weeding, and artificial disturbances).

The contributors of this volume often contrast ADE with the surrounding forest soils based on the assumption that the ADE are anthropogenic and the forest is "natural". What if the benchmark against which ADE is identified and defined is also anthropogenic? If the entire Amazon Basin is to some degree anthropogenic as some historical ecologists argue, the possibility of finding a totally pristine natural soil in Amazônia after thousands of years of human disturbance for comparison as a benchmark is unlikely. The concept of domestication of landscape (Clements et al., 2003; discussed below) may provide an alternative to the site concept.

Most ADE discussed in this volume are entire sites or a subset of traditionally defined sites (most covering hectares). The boundaries of an archaeological site and its ADE do not always correspond such as in the case of the Açutuba site on the Rio Negro where surface artifact scatters much larger than the ADE (Heckenberger et al., 1999) or the Araracuara sites on the Caquetá river where ADE in the form of terra mulata extends far beyond the distribution of artifacts (Mora et al., 1991). How big does a black earth footprint have to be in order to be called ADE. Many archaeological occupation sites have discrete middens that meet the content criteria of ADE; but they are of small-scale contexts within the larger archaeological site (e.g. an individual garbage pit, lens of midden on an abandoned house floor, or post holes packed with dark midden). In addition to color, Kerr et al. (2003) stress that ADE have a greater depth of anthropogenic A horizon than typical forest soils (30-60 cm vs 10-15 cm). Although the cultural strata of most archaeological sites correspond to the modern A Horizons; there are many exceptions such as those that are deeply buried paleosols or are so thick that post-abandonment soil formation processes have not created a deep A Horizon.

THE SIGNIFICANCE OF THE DISTRIBUTION OF AMAZONIAN DARK EARTHS

The total geographical extent remains unknown. Various authors (Kern et al., 2003; Neves et al., 2003; Sombroek et al., 2003) summarize new information about the geographic distribution of ADE in Amazônia from systematic archaeological and pedological surveys. Sombroek et al. (2003) estimate that ADE covers 0.1-0.3% or 6,000-18,000 km^2 of the total Amazon Basin (6 million km2). ADE is reported for most of the major rivers of the Central and Lower Amazon Region. Most are associated with bluffs overlooking várzea near larger active or abandoned river channels and a few have been found on terra firme away from main rivers. Dispersed ADE are reported from certain riverine locations in the Upper Amazon of Peru, Ecuador, and Colombia.

Kern et al. (2003) cite recent surveys documenting ADE "every 5 km along the igarapés, and an over-all spatial coverage of one per 2 km^2". In the Upper Xingú region, Heckenberger (1998) estimates large ring-plaza villages, many of which have ADE, are spaced several kilometers apart. In the Upper Madeira river, ADE is distributed 1 per 2 km^2 (Miller, 1992: 220). High densities of ADE are reported for the Central Amazon Basin (Nimuendajú, 1952; Denevan, 1996, 2001; Heckenberger et al., 1999). Although the best-known ADE are large (ranging from 500 ha for the Santarém site, 200 ha for the Belterra site, 80 ha for the Manacapuru site; and 90 ha for the Altamira site; Smith, 1980, 1999; Denevan,

2001 and authors of this volume), 80 % of the known ADE are less than 2 ha (Kern et al., 2003).

The densest distribution and largest ADE are associated with archaeological settlement sites along the middle and lower courses of the major rivers (Smith, 1980: 562-563; Petersen et al., 2001: 91) or "on the margins or confluence of streams and rivers or near falls" (Kern et al., 2003). As discussed above, in the past scholars have highlighted the distinctions between the potential resources of the larger floodplains (várzea) and interfluvial or upland regions (terra firme) which was reflected in the archaeological record as larger, more permanent settlements and associated with more complex societies established along major rivers and less permanent settlements and simpler societies associated with interfluvial regions. According to these scholars, the large, dense populations of the várzea were sustained by fish and other aquatic resources and farming the rich, annually renewed, floodplain soils of white water (sediment-rich) rivers. Thus, it is not surprising that ADE would be associated with large rivers.

In the 1980s, scholars began to challenge the assumption that the interfluvial uplands (terra firme) were homogeneous and resource poor (Moran, 1991; Viveiros de Castro, 1996; Whitehead, 1996; Heckenberger, 1998, Neves, 1999; Wüst and Barreto, 1999). Thus, finding ADE in terra firme is not surprising (Smith, 1980; 1999; Balée, 1989, 1994; Costa et al., 2003; Myers et al., 2003; Neves et al., 2003). Although inland, most terra firme ADE are on bluffs above smaller upper tributaries and streams. These ADE sites are generally smaller and more dispersed than those

associated with larger rivers (Smith, 1980). The large inland ADE the Belterra Plateau between the Tapajos and Curua rivers and along the Arapiuns river reported in this volume and by Smith (1999) are prominent exceptions. Nimuendajú (1952: 11) reported terra firme ADE associated with deep and wide artificial wells.

In an important revision of the floodplain vs. interfluvial (várzea vs. terra firme) hypothesis about Amazonian cultural development, Denevan (1996, 2001: 102-114; Myers et al., 2003) pointed out that most prehistoric, historic, and present large settlements are located on the terra firme bluffs overlooking an active (or what was at one time) channel of the river rather than on the floodplains of the Amazonian drainage. The Bluff Model highlights the advantages of this location: direct access to floodplain and interfluvial resources, canoe transportation, defense, and dry locations for year-round settlement.

Denevan also points out that most pre-Columbian sites on bluffs are ADE, often surrounded by terra mulata. The large ring plaza villages of the Upper Xingú River are examples of bluff ADE over-looking floodplains in the smaller upper drainages (Heckenberger et al., 1999). As Kern et al. (2003) state, "...ADE are present in practically all types of eco-regions and landscapes," but based on their maps and those of others, the largest and most numerous are found adjacent to lower and central courses of larger rivers (supporting the earlier interpretations of Carneiro, Lathrap, Denevan, Roosevelt, and others regarding the ecological advantages of access to rivers over inland locations for settlement and cultural development).

Returning to distribution of ADE, here we would like to stress two points 1) ADE have not been found everywhere within the Amazon Basin and 2) long-term permanent human settlement does not necessarily result in ADE. ADE are less common the Upper Amazon of Peru, Ecuador and Bolivia, Rio Negro drainage, Orinoco drainage, the Llanos de Mojos of the Bolivian Amazon, and the northern part of the Amazonian drainage basin. Why did ADE form in some areas and not in others? Is the absence of ADE a product of differences in classifications of soils, lack of archaeological and soil survey in these regions, or burial, erosion, leaching, and destruction of ADE sites? Denevan (2001: 114, footnote 4) suggests that lack of reported ADE in the Western Amazon may be due to "lack of awareness".

Only a small sample of the Amazon Basin has been systematically surveyed for archaeological sites; and future research will probably identify new ADE in these regions. On the other hand, many scholars are now familiar with the concept of ADE and actively looking for sites; thus, the geographical distribution may be real and significant. Did these areas originally have ADE that disappeared because of overexploitation, poor management, and/or natural processes of decomposition, erosion and leaching?

Although unlikely, these possibilities should be addressed. An examination of the patterning of spatial and temporal distribution and cultural traditions where ADE is found provide some insights.

Does settlement size and duration play a role in whether or not ADE is formed? Most scholars seem to agree these are important factors. Were the cultures of regions without ADE less evolved and lacking in socio-political complexity compared to those with ADE? A brief examination of two areas that share a similar site plan but not ADE is instructive. The ditched enclosures or ring-ditch sites are elegant patterns of ditches and embankments covering several hectares to several square kilometers located on forest islands and river bluffs in the NE Bolivian Amazon (Denevan, 1966; Arnold and Prettol, 1988; Erickson et al., 1997; Erickson, 2002; Siiriainen et al., 2002). The Bolivian sites are probably part of larger related cultural phenomena of ring plaza settlements associated with complex societies reported for the Upper Xingú, Guaporé/Itenez, and Madeira river drainages (Miller, 1992, 1999; Heckenberger, 1998; Heckenberger et al., 1999). ADE is associated with many of these sites on the Upper Xingú, central Madeira, and Guaporé/Itenez rivers while those of northeastern Bolivia do not have ADE.7 The present day border between Bolivia and Brazil (the Guaporé/Itenez River) may mark a cultural boundary. This also roughly marks the present day transition between high canopy tropical forest and savanna. The Bolivian sites have not been adequately dated but appear to be late prehistoric. The Brazilian sites have a long chronology; and thus, ADE may have formed through longer continuous occupation.

Could the presence and absences of ADE be associated with differences in terms of natural resources, environments, soil types, agricultural practices, or settlement type? These geographical distributions suggest that the differences may be cultural rather than environmental. Basic cultural explanations may account for ADE. Organic matter placed directly in fields and "used up" under cultivation vs. organic matter accumulated to form ADE for later use as farmland are significant farmer decisions that could determine the presence or absence of ADE. In some cases, Amazonian riverine communities dispose of garbage by tossing it into the river. Stocks (1983) reports that the Cocamilla of Peru discard garbage into local lakes that is said to increase the populations of the fish they consume. In contrast, the Tukanoans of the Colombian Amazon (Chernela, 1982) and the Ka'apor of eastern Brazil never dispose of garbage in the rivers (Balée, 1994). A simple decision about garbage may determine whether settlements produce or become ADE or not. In cultures where garbage was tossed into rivers, streams, and lakes, no terra preta-type ADE would form; in others where garbage was deposited and accumulated on or near the settlement, terra preta-type ADE could form.

AMAZONIAN DARK EARTHS AND COMMUNITY SETTLEMENT PATTERNS

Archaeologists have become increasingly aware of the importance of cultural behavior and concepts regarding trash and its patterned, non-random disposal (e.g. Schiffer, 1987). As patterned trash disposal, ADE may provide insights into how people conceptualized and assigned cultural meaning to garbage (trash) and its proper disposal. Native Amazonians are known in the historical and ethnographic literature as "clean" people.

Traditional villages are commonly described as clean, well maintained, and orderly. In the Upper Amazon, native peoples carefully sweep debris on and around house floors and house clearing to the outer edges of the village or hamlet daily (DeBoer and Lathrap, 1979; Zeidler, 1983, 1984; Siegel and Roe, 1986; Stahl and Zeidler, 1990). The cleared area of settlement has rich symbolic meaning as culturally domesticated social space distinguished from the wild, undomesticated space of the forest beyond. A clean village is part of community pride and great effects are made to prepare for important feasts. Clearings provide protection of bare feet against cuts by broken pottery and lithics and snakebite and for removal of decaying organic materials that harbor disease pathogens. Formal garbage disposal was critical for maintaining health in the large populated urban centers of late prehistory.

ADE are highly variable in size, form and depth of deposit. A working assumption held by most of the authors in this volume is that the size and depth of ADE are directly associated with population size of the settlement and settlement duration.8 Another assumption, although less discussed, is that the shape or "footprint" of ADE is associated with community pattern that reflects underlying social organization. Large, multicomponent (occupied for long periods of time by groups of people defined by distinct pottery styles and traditions) sites are found throughout the Amazon Basin regardless of whether ADE is present. Thus large size and long occupation duration are necessary, but not sufficient, explanations for ADE formation. In the following section, we explore two ideas: 1) presence and absence of ADE due to differences in settlement type and garbage disposal patterns, and 2) forms (spatial footprints) and internal heterogeneity of ADE reflect community settlement patterns and/or processes of settlement establishment and reestablishment of settlement space through time.

Could ADE patterns and forms be associated with a certain type of settlement design or village plan? Myers (1973) classified historical and ethnographic accounts of traditional community patterns into "linear" and "non-linear". Linear (one axis considerably longer than the other) includes "lines of houses community" and "multifamily longhouse community". While these basic categories of community pattern account for all ethnographic settlements or archaeological sites, Myers' study presents many valuable testable models.

The "lines of houses community", composed of end-to-end houses parallel to a long plaza overlooking a lake or river, is common in the Napo and Ucayali Rivers of the Peruvian and Ecuadorian Amazon. Garbage is disposed of in middens behind the houses and/or in front of the plaza (DeBoer and Lathrap, 1979; Siegel and Roe, 1986). Myers (1973) attributes the pattern to the exigencies of a linear high ground in the form of narrow levees or bluffs along rivers. The large towns of continuous band of houses for kilometers along the Central and Lower Amazon River reported by early explorers were large versions of the lines of houses community (summarized in Denevan, 1996, 2001; Myers et al., 2003). Apparently many settlements had houses arranged on streets facing public plazas with temples and men's houses. Formal roads leading to the interior are also described. The accounts are not specific about trash disposal patterns but one would assume that the pattern would be wide linear midden(s). The "multifamily longhouse community" or, made up of a single or multiple longhouses (maloca) with up to 100 families (400 people) per house is common in the northwest Amazon (Myers, 1973). The "multifamily long house community" of the northwest Amazon is linear because the domestic structure is much longer than wide (100 m long x 15 m wide; rectangular, oval, or combination in footprint). Garbage is tossed outside the house clearing along the axis of the longhouse.

Myers (1973) provides archaeological examples of linear ADE from the Upper Amazon that may have been formed by linear communities of lines of houses or single long houses. As Smith (1980) points out, the linear ADE in the Central and Lower Amazon extend hundreds of meters back from the bluff edge; thus, these communities must have had multiple rows of lines of houses. Many of these settlements may have been associated with the historical Tapajós chiefdom and archaeological ADE of a predominately linear type (Nimuendajú, 1952; Denevan, 1996, 2001). Nimuendajú (1952) reports 65 Tapajós sites most of which are ADE and estimates that the total number is probably double.

Myers' non-linear pattern (circular, oval or amorphous in footprint) ranges from the "isolated single family house community" to the large "central plaza type community". I also add an intermediate type: the "multifamily roundhouse community" found in the Orinoco River drainage (Wilbert, 1981) and the "house lot community" based (discussed below). The "isolated single family house community" and the "house lot community" are common throughout Amazônia today. A single extended family house is surrounded by a 30-m diameter cleared area with a shallow "doughnut-shaped" ring of refuse around the clearing created through daily sweeping (DeBoer and Lathrap, 1979; Zeidler, 1983, 1984; Stahl and Zeidler, 1990). Another hypothetical disposal pattern would be accumulation of garbage into a single heap. Refuse also accumulates under the house on the house floor, especially near cooking hearths (DeBoer and Lathrap, 1979).

A larger scale version of the above is the "multifamily roundhouse community" in which all reside within a single structure of 15 m diameter (Wilbert, 1981). The pattern of garbage disposal is expected to form a large doughnut shaped ring around the house and public plaza cleared area.

"Multiple longhouse communities" of the northwest Amazon and "multiple roundhouse communities" of northern South America are made up of various multifamily houses that together produce non-linear settlement pattern. Some of the larger communities are arranged around plazas. Garbage is either swept into a ring beyond the clearing of the cluster of long houses or in rings around each individual house. I would also add the "house lot community" based on traditional Maya urbanism (Killian, 1992). These are characterized by relatively regularly dispersed independent households with patterned spatial organization of house, outdoor activity areas, midden, and gardens.

The "central plaza-type community" is the most highly structured, non-linear community (Myers, 1973). Those described for the Bororo, Gê, and Carib-speaking peoples in Central Brazil are made up of a circular ring (or concentric rings) of up to 8-31 houses around a circular plaza of 110-300 m diameter. Communities of over 140 houses for up to 1600 people arranged in 3 rings around a plaza are documented (Nimuendajú, 1946; Bennett, 1948; Myers, 1973; Wüst 1994; Wüst and Barreto , 1999).9 Garbage is placed in individual piles up to 10 m "behind" the houses that face the plaza (ibid.; Heckenberger et al., 1999). Over time, the piles of midden form a doughnut shaped ring beyond the house circle creating what would be a 350 m diameter or 10 ha site (Myers, 1973).

Myers predicts that the sweeping effect in the plaza and the mounding effect in the midden would create slight topographic differences. Variations include a central plaza-type community with a square or rectangular plaza and a street like arrangement of 4-8 multifamily longhouses each holding 30-200 families described for the Tupinamba of the Brazilian coast village of 4-8 houses, each housing 30-200 families (Lowie, 1948: 16). Colonial documents for Bolivia report large towns and villages of thousands of inhabitants with central plazas with up to 400 houses on streets, presumably organized as a rectangular or square grid (Denevan, 1966). The distribution of midden would probably be similar to that of the central plaza-type community or the house lot community. Myers (1973) notes that central plaza-type communities would be much easier to defend using ditches and palisades than linear villages.

Most small, dispersed shallow lenses of ADE on terra firme are probably isolated single family house communities but with a center type ADE instead of the doughnut shaped ring. The oval and elliptical mound ADE of the

Marajoara culture on Marajó Island (Roosevelt, 1991) and the Bolivian Amazon (Denevan, 1966; Erickson, 2000b; Langstroth 1996) probably represent single or multiple long house communities (Myers, 1973). Archaeologically, the central plaza-type community includes the large central plaza sites often with arcs, circles, or rectangles of ditches and embankments described for the Upper Xingú Basin (Heckenberger, 1998; Heckenberger et al., 1999), Upper Madeira Basin (Miller, 1992, 1999), the Bolivian Amazon (Arnold and Prettol, 1988; Erickson et al., 1997; Siiriainen et al., 2002), the Açutuba and Osvaldo sites on the Rio Negro (Heckenberger et al., 1999), and ring plaza villages without earthworks for the Tocantins River Basin (Wüst 1994; Baretto and Wüst, 1999).

The oldest central plaza-type community archaeological sites are those of the Valdivia Culture in Ecuador (3500-1500 BC) with a large rectangular plaza surrounded by a ring of densely clustered multifamily houses (Lathrap et al., 1977; Stahl, 1984; Zeidler, 1984; Lathrap, 1985; Lathrap et al., 1985; Raymond, 1993).10 The majority of the central plaza-type community archaeological sites discussed above are concentric type ADE, in some cases with multiple rings of ADE such as on the Tocantins River (Wüst and Barreto, 1999). The primary settlement of the late prehistoric Tapajós culture is under the present day city of Santarém. The ADE is up to 1.5 m deep and estimated to cover 4-5 km2 and assumed to be non-linear (Nimuendajú, 1952: 9; Smith, 1980; 1999; Roosevelt, 1999: 24).

Although both linear and non-linear community patterning are documented, Smith (1980) and Denevan (2001: 105) stress that most known ADE are linear (one axis longer than the other) and laid out parallel to the bluff edge and/or nearest active or once active river channel. The long axis of many of the larger ADE extends several kilometers (Smith, 1980; Denevan, 2001: 105); thus, we may be seeing the accumulated result of many communities established and reestablished along the bluff over thousands of years.

Another distinction raised in this volume is "center type" and "concentric type" ADE that may also reflect community patterning and complex site formation processes. Sombroek et al. (2003) characterize the concentric type ADE as having a deep concentric ring(s) of ADE around a relatively clean central area (the classic doughnut shaped midden discussed above) and the center type ADE in which the deepest anthropogenic soil is in the center and tapers off towards the edges of the site (mound-like midden). The concentric type ADE (e.g. best documented archaeologically in the Tocantins and Upper Xingú river basins by Wüst and Barreto, 1999) is assumed to correspond to the single house community and multiple family roundhouse community if small, and the central plaza community and its variants if large.

The majority of ADE are center type: they have continuous distribution of black earth across the site and are deepest in the center with no evidence of an ADE-free central plazas (Sombroek et al., 2003). The center type ADE does not closely fit any of the community patterns discussed above. Does this mean that pre-Columbian community was of a form that is not represented in the ethnographic and historical literature or could it be the result of complex site formation processes? The center type ADE implies that 1) people lived on top of their garbage (e.g. the moundbuilder model where trash and fill are used to raise the settlement; possibly for visibility,

drainage, or an expression of monumentality); 2) garbage disposal for fields was spatially discrete from residence (e.g. Kayapó enriched garden model); 3) each community maintained a single location to pile garbage (rather than sweep it into a doughnut shaped ring around the residential clearing); and/or 4) slight shifts of the community location or residences within the community over time distributed the midden across the entire site.

Sombroek et al. (2003) and Denevan (2001) suggest that ADE may have started off as a concentric type but became center type due to a "smearing effect" caused by periodic reestablishment of the community through slight movement of the center of the settlement. Scholars have documented this phenomenon for Xinguano, Gê, and Bororo communities (Heckenberger et al., 1999; Wüst and Barreto, 1999). In most cases, the community is moved several hundred meters because of "rotting houses, frequent deaths, internal disputes, warfare, and sanitary conditions" (Wüst and Barreto, 1999: 12).

These hypotheses could be tested archaeologically with careful horizontal areal excavations and sampling of larger sites and ethnographic contexts. Archaeologists have shown in ethnographic and archaeological cases that despite careful daily sweeping, some garbage remains on house floors due to the "trampling effect" (DeBoer and Lathrap, 1979; Zeidler, 1983, 1984; Siegel and Roe, 1986; Stahl and Zeidler, 1990; DeBoer, 1996); thus, the house clearing could accumulate continuous ADE given enough time.

The buildup of midden on house floors and under storage racks over several years can be quite substantial. Few studies of differentiation of cultural space and heterogeneity of ADE through mapping and excavation have been done. Nimuendajú (1952: 11) pointed out, "The surface of the black earth deposits is usually not flat but composed of a number of mounds, each several meters in diameter, and each probably representing a house site".

In this case, the uneven surface of ADE may reflect the differential use of cultural space within the site. Heckenberger et al. (1999) detected a large rectangular plaza surrounded by artificial temple and/or elite residential mounds through surface topography, ditches, thickness of ADE, and distribution of artifacts at the Açutuba site.

In most ethnographic cases discussed above, garbage middens and domestic zones within the settlement are spatially discrete (garbage is swept or tossed beyond the house clearing or central plaza house circle clearing). If all organic matter and charcoal generated by the community were systematically gathered to create ADE outside the residential zone in piles, ring, or arc of midden, or directly in fields, there should be a large, relatively organic-free zone in the residential sector of each site, no matter what type of community pattern. Center type ADE probably forms where sites have been continuously occupied for long periods of time. In large pre-Columbian settlements, it may not have been possible to dispose garbage far from the residence because of close neighbors; thus midden may have accumulated around and under houses as predicted in the house lot community or multiple reestablishments of any community pattern in the same general location.

Later inhabitants of the settlement would have to periodically decide whether to continue piling up new trash on the already established piles or rings inherited from their ancestors (maintaining a concentric type ADE) or to level the surface by filling in low spots with new trash (creating a center type ADE). In my experience digging archaeological sites in North and South America, organic matter, potsherds, and construction debris were used to level low spots, fill unused pits, and abandoned structures producing more dispersed distribution of localized ADE (creating greasy, black soils with considerable charcoal, ash, bone, and shell, what archaeologists usually referred to as "dark midden").

Most Amazonian peoples raise their house floors for improved drainage. Once organic matter has decomposed into mature ADE and not considered a health and aesthetic problem, it can be lived on or treated as transportable construction fill and fertilizer for agriculture. In this scenario, people literally lived on their garbage but only after it had been converted into harmless ADE.

The presence of the house garden may have been a determining factor in formation of the center type ADE. Many authors of this volume (Clement et al., 2003; Hecht, 2003; Hiraoka et al., 2003; Silva, 2003) discuss the importance of the house garden as the prototype for the formation of ADE. The gardens that regularly receive organic matter from food preparation and cooking are located near the kitchen (thus residence and garden spaces are in close proximity).

In small settlements the garden is always adjoining the domestic space. As settlement size is increased, gardens can either be relocated outside the residential area or squeezed into spaces between houses (house lot community). There were fewer domestic animals (such as introduced pigs, goats, and chickens) to compete for house garden space in pre-Columbian settlements, thus house gardens may have been larger and more common than in modern villages. Over time, this strategy would generate a center type ADE.

AMAZONIAN DARK EARTH FORMATION PROCESSES

Although Amazonian soils are not as poor as presented in the traditional literature, ADE clearly stand out as anomalous. The authors of this volume agree that ADE are human produced or anthropogenic. Classic ADE from the Upper Amazon to the mouth of the Amazon are archaeological sites.12 The terra mulata soils are also clearly human altered landscapes because most are associated with terra preta and archaeological sites. Because pottery, bones, shell, and other domestic debris is found throughout the soil profiles, ADE are sites of human occupation and probably farming rather than simply locations of naturally rich soil that were colonized for residential and agricultural use.

The authors concerned with the origins of ADE are interested in what archaeologists and geomorphologists since the 1970s have called "site formation processes" which include both natural and cultural transformations (Schiffer, 1987; Ruivo et al., 2003; Silva, 2003) or "middle range theory" (Binford, 1978). Kämpf et al. (2003) distinguish between "product" and "process" in ADE research. The product is relatively well known and agreed upon; the process is still in debate. The contributors of this volume agree that ADE are anthropogenic (associated with human settlement and/or farming) rather than natural. Most agree that the primary element in ADE formation is the concentrated accumulation of organic matter created through permanent settlement and farming in large farming communities.

ADE formation must be the result of a significant shift in human behavior and activity during late prehistory in the Amazon Basin. The distribution of ADE is "not associated with a particular parent soil type or environmental condition" (Kern et al., 2003), thus ruling out some environmental determinants.13 Neves et al. (2003) and Petersen et al. (2001) state that the "basic precondition" for the formation of ADE is a lifeway based on sedentary villages and intensive agriculture. They argue that sedentism was a late phenomenon in the Amazon appearing around 2000 years ago. Most archaeologists accept the idea that sedentism began thousands of years before the widespread appearance of ADE (Lathrap, 1977; Roosevelt, 1991; Roosevelt et al., 1991), thus sedentism alone cannot account for the formation of ADE. The dates for raised fields throughout the Neotropics support the hypothesis of Petersen et al. that intensive agriculture was adopted between 1000 and 2000 years ago (Denevan, 2001).

Kämpf et al. (2003) divide the processes of formation of ADE and other anthrosols in the Amazon into three overlapping groups 1) "midden model" (Smith, 1980) based on "unintentional anthropic activities" related to permanent settlements and the accumulation of domestic refuse; 2) "agricultural model" (Denevan, 1998; Woods and McCann, 1999) based on "intentional anthropogenic activities" related to semi-permanent agriculture involving clearing and burning of vegetation; and 3) "moundbuilder model" based on "intentional anthropogenic activities" related to transformation of soils for raised fields and mounds. In agreement with most other scholars, Kämpf et al. (2003) attribute terra preta ADE to the midden model and terra mulata ADE to the agricultural model. I disagree with their conclusion that the formation of terra preta-type ADE was unintentional.

All the contributors to this volume agree that domestic refuse generated in human settlements is the key component of ADE. Human refuse is assumed to include general kitchen debris, food remains, food processing wastes, feces, urine, house construction materials, pottery, lithics, shell, ash, and charcoal. As Costa et al. (2003) stress, the bulk of the organic refuse has decomposed or weathered beyond recognition or is lost through leaching and cultivation; thus, it is difficult to identify the origin and the percentage of contribution to ADE. Pottery, bone, shell, plant remains, construction materials, fuel, and lithics can be identified and quantified, although the relative durability of certain materials may bias our understanding of ADE. The authors of this volume use a variety of techniques for determining ADE components.

Most authors agree that the high percentage of charcoal (pyrogenic carbon or black carbon) that makes up over 1% of total mass distinguishes ADE from other anthropogenic or "natural" soils (Glaser et al., 2003; Kämpf et al., 2003; Kern et al., 2003; Sombroek et al., 2003; Thies and Suzuki, 2003). According to Glaser et al. (2003), ADE had 64 times more charcoal per unit volume than adjacent forest soils. Charcoal not only helps maintain fertility, but also is an important carbon sink. The authors attribute the stability of the organic component, and hence high soil fertility to the black carbon. If carbon is the key, the producers of ADE may have employed a special kind of burning or treatment of disposal of charcoal and ash that was distinct from Amazonian groups that created non-ADE anthropogenic soil under practicing similar contexts of settlement and agriculture. We can also infer that fuel must have been abundant and readily available because of the quantity of intact pieces of charcoal rather than ash. If fuel was scarce, we would expect more complete burning or reburning of any remaining charcoal from earlier fires reduced it to ash. This is remarkable given the large and dense populations living in pre-Columbian settlements and their demands for fuel.

I agree with Sombroek et al. (2003) that a higher level of charcoal in ADE than in surrounding forests is evidence of intentional management (contradicting the claim by Kern et al. (2003) that ADE formed under the Midden Model was "unintentional"). Native Amazonians apparently controlled the burning process in such a way that charcoal rather than ash was produced. Thus, these practices were important in the formation process of both the terra preta and terra mulata ADE.

This specific burning was distinct from natural forest fires that periodically rage through Amazônia, burns used in swidden agriculture, and domestic fires in households of non-ADE sites or modern settlements (Glaser et al., 2003). Hecht (2003) suggests that the controlled "low biomass cool burn" of crop residues, weeds, and palm fronds practiced by the Kayapó combined with charcoal from cooking hearths could account for the charcoal in ADE. Anyone travelling through the Neotropics cannot fail to notice the ubiquitous smoky and smoldering fires found in rural and urban contexts. Amazonian peoples use regular low intensity burns to dispose of household garbage, weeds, and crop debris. In addition to trash removal, the smoke generated helps to keep insects at bay.

Most authors of this volume are also in agreement that carbon within ADE created a favorable habitat for beneficial microorganisms that may be the key to the creation and maintenance of high soil fertility (Woods and McCann, 1999; Falcão et al., 2003; Kern et al., 2003; Sombroek et al., 2003). Thies and

Suzuki (2003) refer to studies that "...soils containing high levels of BC (black carbon) have correspondingly higher levels of microbial biomass and respiratory activity" and that "Amazonian Dark Earths are likely to contain several billion viable microorganisms per gram, representing many thousands of different species". Lehmann et al. (2003b) show that the conditions for biological nitrogen fixation are optimized in ADE.

The authors are in general agreement that population density and duration are important factors in ADE formation; but few would concur that this is an adequate explanation. Most communities between AD 1 and 1500 were producing the same classes of domestic refuse, and large sedentary communities were widespread throughout the Amazon. Other than the frequency of charcoal, what was unique about the type or composition of refuse produced by dwellers of what would become ADE? There are few systematic studies to quantify the amount and type of domestic refuse produced by native Amazonians. I will briefly examine four classes of refuse that have not been adequately considered in this volume: fish and fishing, pottery, roofing thatch, and night soil.

Fish is a class of domestic refuse that may have played a key role in the formation of ADE. The debates about protein availability as an environmental determinant of cultural development in the Amazon in the 1970s highlighted the importance of fish in the diet of Amazonian societies (e.g., Lathrap, 1970; Beckerman, 1979; Roosevelt, 1980; Carneiro, 1996; Gragson, 1992; Erickson, 2000a). Many now agree that fishing (including shell fish such as gastropods and molluscs), combined with plant sources such as palm and fruit trees, provided much more of the protein for Native Amazonians than hunting. Could fish and fishing also contribute to the formation of ADE? As every author in this volume points out, most ADE are found on terra firme overlooking rivers, lakes, floodplains, and wetlands (Smith, 1980; Denevan, 2001; Kern et al., 2003). Kern et al. (2003) state, "Maps produced by archaeologists with the distribution of archaeological sites, show a predominance of sites placed on the margins or confluence of streams and rivers or near falls". It is well known that these are the best locations to fish in the Amazon. Could the source of nutrients so important for establishing and maintaining ADE discussed by Glaser et al. (2003) be remains of the processing, preparation, and the consumption of fish?

The famous Araracuara ADE site in the Colombian Amazon is located on an unnavigable stretch of the Caquetá River where the settlement historically has been associated with portage of canoes, temporary camps of fishermen and travelers, and fishing. Archaeologist Agusto Oyuela (pers. comm.) has found ADE in similar locations along other rivers of the region. This hypothesis for ADE formation could be tested through archaeological recovery of the more durable parts of fish (scales and ooliths), ethnographic observation, controlled experimentation, and the identification of relevant molecular biomarkers (Woods personal communication). Sombroek et al. (2003) briefly mention the possible contribution of phosphate-calcium compounds from fishing and hunting and calcium from gastropods and molluscs. More recently, Lehmann et al. (2003a) attribute the high levels of phosphorus and the presence of combined calcium-phosphorus in ADE to massive accumulation of fish waste.

Most authors note that broken pottery (potsherds, sherds) and fired clay probably from dismantled fire hearths and burned wattle and daub structures are a major component of terra preta type ADE (Costa et al., 2003; Myers et al., 2003; Neves et al., 2003). Could fired clay in its various forms have contributed to the creation and maintenance of ADE? High frequencies of potsherds in the soil profile could affect drainage, texture, chemical and biological activity. Organic tempering materials (ash, shell, sponges, and Spanish moss) used in Amazonian pottery (identified by Costa et al., 2003) may also contribute to the formation of ADE.

Pottery is so common in some pre-Columbian occupation mounds and forest island sites in the Bolivian Amazon that Robert Langstroth (1996) has coined the term "sherd soils" to describe them. Reading the methods sections of the laboratory analyses of ADE in this volume, I learned that pottery is systematically removed from soils during processing for soil analysis (e.g. Teixeira and Coimbra, 2003). I suggest that sherds should be included in soil samples and analyzed to determine their potential contribution to ADE.

Herbert Smith (Smith 1879 cited in Woods, 2003) suggested a connection between palm thatch and ADE in 1879. Gragson (1995) also suggests that economic palms contributed to ADE. He demonstrates that the thatch roofs of a typical Pumé community require 13,498 fronds of Mauritia flexuosa, which contains 3,373 kg of dry matter. Considering that thatch is replaced every 2-3 years and house poles of palm wood at somewhat longer intervals, the contribution of organic matter to settlements could be substantial. When these figures are applied to the hundreds of large houses for towns described in early historical accounts before massive depopulation, the amount of organic matter is staggering. Because palm fronds are transportable on the backs of humans or by canoe, this may be an important non-local source of organic matter. Sombroek et al. (2003) note that palms, possibly used as house thatch, have been identified in pollen analyses of ADE.

Woods (2003) briefly mentions night soil (human feces and urine) as contributing to phosphates in ADE. Systematic manuring using animal dung was a key component of the Plaggen soils of Europe. In contrast, Amazonian peoples had few domestic animals (muscovy duck, dog, and keeping of wild tamed animals and birds). Carvajal describes huge complexes of turtle pens in some of the towns he visited, a possible source of considerable organic matter. Human burials may have also contributed to ADE, especially after population density in settlements increased. Traditionally, Amazonian peoples buried their dead below house floors, in plazas, or in separate cemeteries. In some cases, bodies were cremated.

The authors of this volume suggest that some forms of ADE (terra mulata) may have been the long-term result of improving cultivated soils with organic matter through burning, mulching, composting, and fertilizing. Farmers may have done these activities within the fields, through transportation of mature organic matter from residences and from forest to fields, or some combination.

Most authors agree that carbon, in the form of charcoal, is an important element in ADE and may provide the explanation for the resilience of ADE to degradation and erosion and its ability to support microorganisms. The soils associated with swidden agriculture have been well studied in the Amazon and elsewhere.

Although swidden is regularly practiced on ADE and produces charcoal (German, 2003), swidden agriculture does not produce ADE. Many scholars (Smith, 1980; Eden et al., 1989; Denevan, 2001; Myers et al., 2003; Sombroek et al., 2003) argue convincingly that swidden agriculture did not contribute to ADE nor was it an important farming strategy before the introduction of the iron and steel ax and machete.16 Today, swidden farmers in the Bolivian Amazon generally burn and reburn (often numerous times) the fallen trees and debris from field clearing and preparation. Most tree trunks that do not burn are eventually cut up and used for fuel in the village or left to rot. Fires used to cook and provide warmth at night generally burn continuously, producing more ash than charcoal.

To intentionally increase levels of charcoal in the soils, native Amazonians may have practiced a special kind of burning under controlled conditions or what Sombroek et al. (2003) refer to as "controlled carbonization". One possibility is that they used a reduced atmosphere burning similar to that employed to produce charcoal. The source of the charcoal in ADE is assumed to be from both primary forest and secondary forest around the settlement. Could the charcoal that formed ADE have been primarily from certain plants whose burning is more amenable to charcoal production? Experiments, ethnographic observations, and archaeobotanical analysis of fuel from ADE may provide clues to the sources (e.g. Mora et al., 1991; Piperno and Pearsall, 1998; Mora, 2003).

Controlled burning, and thus production of charcoal and ash, may have been more common in pre-Columbian Amazonian settlements than today. Large populations would have greater needs for high-quality fuel for cooking; smoke-preservation of foods; conversion of bitter manioc into storable flour and breads; salt making; pottery firing; and other domestic activities. As forests around the settlements were gradually turned into gardens, fields and orchards, firewood would be imported from longer distances (using water transportation and capture of driftwood). As settlements grew, ritual and feasting activities became more common and larger events consuming additional fuel for food preparation and pottery production (Neves et al., 2003). The technique of reduced atmosphere pottery firing is well documented in the archaeological record. Does this imply knowledge of charcoal making was employed in ADE production?

As Woods (2003) notes, the rate of deposition of organic matter must be greater than losses in order for ADE to form. Natural processes that contribute to losses included volatilization, leaching, and erosion "which are expected to be high in a hot, high rainfall and humid environment such as Amazônia". One means of effectively raising the ratio of inputs to losses would be the concentration of nutrients from a larger region to a small concentrated area that becomes ADE. This implies efficient transportation (canoes, canals, paths, roads) and organization of labor. Hunting, gathering, fishing, and offsite farming are typical activities that concentrate non-local organic matter in settlements.

We would add the losses due to crops extracting nutrients from farmed ADE. The balance between rate of deposition and loss highlights the inherent conflict between the simultaneous formation and maintenance of ADE through domestic activities and its use for agriculture. In order to fully exploit ADE for cultivation, farmers had to either inherit already formed ADE or wait many years before reaping the benefits of the ADE they produced. Based on archaeobotanical and sediment analysis, Mora et al. (1991: 77) document simultaneous use of ADE in Araracuara for settlement and farming of field crops and fruit trees.

This temporal dimension of ADE formation has important implications. Farmers who created ADE through considerable labor had to wait for a period of time before being able to exploit their creation for farming. If farmers were inverting organic matter into fields that were immediately cultivated (as is the case of the Kayapó), the ratio of nutrient accumulation to losses for ADE formation and maintenance may have been disrupted. If continuous cropping was combined with regular nutrient additions and enrichment to the soils, organic matter accumulation may have offset the losses through use.

TEMPORAL DIMENSION OF AMAZONIAN DARK EARTHS

The formation processes of ADE include a temporal component. The depths of more than 2 m of ADE (average depth

0.73 m) suggests that they were formed over long periods of time (Smith, 1980). The authors of this volume agree that ADE could be formed by large groups of people over hundreds of years. Could ADE also be created by small groups of people over longer periods of time? Although the two scenarios may produce the same result, the archaeological signatures would differ. Based on Woods' insight, middens produced by smaller groups over long periods of time would be subject to higher nutrient losses through natural processes, possibly off-setting accumulation of organic matter. The number of sites covering a hectare or less demonstrate that small groups did create ADE.

Smith (1980: 564) estimates an accumulation rate of 1 cm per 10 years (or 10 cm per 100 years). Smith (citing C. Evans, 1964) also reports 1 m of ADE forming at sites on Marajó Island within 100 years (1 cm per year), but discounts the rate as impossible. For the ADE of the Lago Grande site, Neves et al. (2003) cite a figure of "a hundred years for forty centimeters of terra preta" or 0.4 cm per year.

Using dates provided by Petersen et al. (2001: 97-101) and Heckenberger et al. (1999: 359, 367), I calculated rates of accumulation required to produce deep ADE. The 30-80 cm deep ADE at the 30-90 ha Açutuba site formed over 1800 years (360 BC and AD 1440) or 0.002 to 0.004 cm per year accumulation. In the Upper Xingú region, a depth of 40-50 cm of ADE in sites covering 0.3-0.5 km2 formed over 700 years (AD 1000 and AD 1700) or 0.06 to 0.07 cm per year accumulation. The 75 to 90 cm [deepest] of ADE in the 32-ha Araracuara site formed over 900 years, or at a rate of 0.08 to 0.1 cm per year (Mora et al., 1991). Although these figures are crude estimates, the rates suggest that ADE formation is slow.

INTENTIONALITY AND AMAZONIAN DARK EARTHS

All authors of this volume agree that long-term permanent settlement in the same location is a key component of formation process of terra preta-type ADE, but not necessarily a complete explanation. Did Amazonian peoples intentionally create and maintain both terra preta and terra mulata-types of ADE for farming or was it the unintentional by-product of permanent settlement and semi-permanent cultivation? All authors of this volume agree that of terra preta-type ADE was the unintentional by-produce of long-term permanent settlement in the same location.

The hundreds of years necessary to produce mature ADE also suggests that ADE formation was an unintentional process spaning many human generations. In contrast, most scholars believe that the formation of terra mulata-type ADE which implies intentional, possibly labor-intensive movement of organic matter from domestic contexts to fields, cutting and removing trees with stone axes, and systematic burning as part of semi-permanent agricultural activities (Sombroek 1966: 175; Andrade, 1986; Mora et al., 1991; Woods and McCann, 1999; Denevan, 2001; Mora, 2003; Myers et al., 2003; Kern et al., 2003).

As an historical ecologist, I am convinced that Amazonian peoples knew what they were doing in regard to settlement, agriculture and other activities that involved environmental transformations over the short and long term. Amazonian farmers today are certainly knowledgeable about the soils that they chose to cultivate and we can assume that was the case in the past. Terra preta-type ADE probably began to form under conditions of permanent settlement. Settlements are the result of cultural decision-making and choices of settlement location and longevity are intentional, conscious acts. Based on excavations of pre-Columbian settlements and observations in historical, ethnographic, and contemporary contexts, we assume that patterned treatment and disposal of trash is also cultural and intentional. But was the intent to create ADE?

The creation of midden that resulted in ADE is primarily a cultural formation process (although natural processes of decomposition, weathering, and bioturbation certainly played a role). Farming is another process. Most Amazonian houses past and present had adjacent house gardens (Lathrap, 1977).

The house garden is a form of intensive agriculture that receives kitchen wastes and is the primary midden in Amazonian settlements (Clement et al., 2003; Hecht, 2003; Hiraoka et al., 2003). If the ethnographic cases are relevant to the past, ADE was probably exploited for cultivation both during and after its formation.

RELATIONSHIP OF AMAZONIAN DARK EARTHS TO PRE-COLUMBIAN SOCIETAL COMPLEXITY

Large scale ADE first appears between 450 BC and AD 950 (Petersen et al., 2001: 100-101; Neves et al., 2003). Petersen et al. (2001: 100) state that the appearance of ADE is "broadly synchronous" throughout the Amazon Basin, although 1400 years is a considerable span of time. They note that in most cases, the earliest dates are from the Central and Lower Amazon associated with the Amazonian Polychrome tradition (in agreement with arguments about the geography and timing of Amazonian cultural development (Lathrap 1970; Lathrap et al. 1985; Brochado 1984)).

What was the unique "kick" or combination of cultural, economic, technological, social or political factors that triggered ADE formation on a vast scale throughout the Central and Lower Amazon in late prehistory? Neves et al. (2003) and Petersen et al. (2001) stress sedentism as the motivating factor ("long-term intensive settlement"). They argue that manioc and intensive exploitation of riverine resources made sedentism possible. This is an unsatisfactory explanation because sedentism, manioc, and aquatic resource use appeared thousands of years earlier in the Amazon (Roosevelt et al., 1991; Piperno and Pearsall, 1998; Oliver 2001). Lathrap (1970, 1977) suggested that the domestication of bitter manioc that could be detoxified to produce a storage product created the context for population growth, larger settlements, warfare, and more complex socio-political organization in the millennium before the appearance of ADE. Roosevelt (1980, 1991) has argued that it was the adoption of maize as a storable protein resource around 2000 years ago that triggered socio-political complexity in the Central and Lower Amazon. Maize was the staple of the Tapajós (Nimuendajú, 1952: 14, Footnote 5) while manioc was the staple of cultures in the Upper Xingú region. Both are late prehistoric inhabitants of ADE sites.

As many archaeologists have noted, the period when ADE appears in the archaeological record is characterized by fancy polychrome pottery, larger vessels, elaborately carved stone objects, and increased local and regional interaction through trade, exchange, feasting, warfare and alliances (Lathrap, 1970; Lathrap et al., 1985; Roosevelt, 1991; Heckenberger et al., 1999; Petersen et al., 2001; Myers et al., 2003; Neves et al., 2003). Whatever the trigger(s), most scholars agree that these changes were associated with population growth and subsequent population pressure in many diverse ecological contexts in Amazônia and increased social interaction (Lathrap, 1970; Roosevelt, 1980; Brochado, 1984; Carneiro, 1996; Oliver, 2001; Peterson et al., 2001).

Was maintaining residence in the same place for many, possibly hundreds of, successive generations necessary to produce ADE?17 Factors range from environmental (desire to control prime real estate with access to multiple resources, river transportation, and farmland; Denevan's Bluff model) to cultural (a strong sense of place and identity), and political (territoriality driven by warfare, competition, and alliances) (Carneiro, 1970; Lathrap, 1970; Denevan, 1996, 2001). Late prehistoric Amazonian warfare may have been a factor in settlement location, longevity, and community pattern with a selection for centralized, densely packed settlements on river bluffs that could be defended with earthworks and palisades (Carneiro, 1970; Lathrap, 1970; Heckenberger, 1998).

The new sociopolitical systems documented at regional scale permitted huge populations to occupy larger settlements than were possible under less complex institutions (Heckenberger, 1998; Heckenberger et al., 1999; Neves et al., 2003).

Because large dense populations are often markers of socio-political complexity, are ADE necessarily associated with or the product of centralized complex societies? Neves et al. (2003) correlate ADE with cultural phenomena appearing 2000 years ago such as earthmoving, long distance trade, impressive polychrome pottery styles, intensification of agriculture, sedentism, roads, and plaza ring ditch settlements which imply increasing hierarchical, socio-political organization. Mora (2003) directly associates ADE with the appearance of chiefdoms in the Upper Amazon Basin. Although many Amazonian societies were probably hierarchically organized and complex by late prehistory, the causal relationship between complex society and ADE is poorly demonstrated.

Other questions about the necessary association of ADE and socio-political complexity can be raised. Kämpf et al. (2003, citing Miller, 1999) report ADE dating to 4,800 years BP, much earlier than any evidence for chiefdoms and states in the Amazon. In addition, if site size is associated with socio-political complexity, it is worth examining the entire range of ADE. Although emphasis in this volume has been on the larger ADE, Kern et al. (2003) point out that most ADE are small (80 % of the known ADE sites cover less than 2 hectares.). Archaeologically, small ADE probably represent individual homesteads, hamlets and small villages of farming societies (Myers, 1973; Smith, 1980). Whether these were the lower rungs of a regional, hierarchically organized socio-political organization has not been adequately determined. Although the cultural phenomena that sparked ADE formation may be associated with sociopolitical complexity, these small sites show that small communities were capable of creating ADE. Density of settlement population was probably more important than total number in the formation of local ADE.

The labor, energy, and costs involved in the creation, use, and maintenance of ADE are also unknown. If the estimates of Mora et al. (1991) are correct, high, sustained labor input was necessary to create the extensive ADE documented throughout the Amazon Basin (discussed below). Was ADE produced by small groups of people over long periods of time or large groups over short periods? Does ADE formation and management require complex social organization or could small autonomous groups have done it? The variation in size and depth of ADE suggests that it was associated with dispersed households, hamlets, villages, and large urban settlements.

AMAZONIAN DARK EARTH AS RESOURCE MANAGEMENT

We see the end result of ADE formation in the archaeological record and soil profiles. The process that produced ADE must be inferred through multiple lines of evidence. I believe that Amazonian farmers would recognize the benefits of ADE as it was forming. If so, both terra preta-and terra mulata-types of ADE were created through intentional and deliberate activities, possibly even involving planning and design rather than unintentional by-products of long-term settlement. As such, ADE are part of a lost indigenous knowledge system, a form of resource creation and management governed by a set of cultural rules, logic, knowledge, and behavior. Until we know more about how ADE was created, it is difficult to address issues of resource management, appropriate technology, and sustainability. Experiments, ethnographic analogy, and cross-cultural analogy can help revive the lost indigenous knowledge system. Insights provided by Hecht (2003), Posey (1998) and WinklerPrins (2001) are important advances.

Is ADE a sustainable resource management system? The fact that (1) ADE are well preserved, stable, and resistant to leaching, erosion, and mineralization (possibly even "growing" as suggested by some scholars) after 400 to 500 years of neglect and/or use, and (2) archaeological dating demonstrates that many ADE were occupied for 900 years or more (Smith, 1980; Denevan, 2002; Kern et al., 2003; Lehmann et al., 2003a; Mora, 2003; Neves et al., 2003) are indications that ADE are sustainable (or at least the human lifeway that produced them was sustainable). The degradation, destruction and loss of ADE today is due to physical removal of the sites, not overuse or declining soil fertility. Abandonment of ADE used for farming today appears to be due to the increase in labor dedicated to weeding (weed invasion) rather than soil nutrient depletion through overuse (German, 2003; Major et al., 2003). ADE are remarkably resistant and resilient.

AMAZONIAN DARK EARTHS, BIODIVERSITY, AND AGRODIVERSITY

The relationship of ADE to biodiversity and agrodiversity is poorly understood. Some authors (Clement et al., 2003; German, 2003; Major et al., 2003) have proposed that because of their high soil fertility, ADE could be rich reserves of

biodiversity, at least in terms of economically useful species for humans past and present. One could also infer from this that game animals hunted by Amazonian peoples are also more abundant because of ADE (since they tend to feed on the same economic species as humans). Balée (1989, 1994; Balée and Campbell, 1990) provides an excellent analysis of the economic species, the floral and archaeological signature, and biodiversity associated with ADE in central Brazil. Historical ecologists point out that the combination of human-enhanced soil fertility, continual disturbance, and selection for economic species can raise biodiversity and increase biomass. The weedy species, feral crops, and semi-domesticated trees commonly associated with long-term human activities and disturbance (especially old paths, fields and settlements) are important elements of agrodiversity and valuable in situ genetic reserves (Balée, 1994; Brookfield, 2001; Clement et al., 2003; Major et al., 2003).

On the other hand, more human interference is often equated with deforestation, excessive disturbance, and creation of secondary forest and grasslands, changes that are often attributed by natural scientists as detrimental to Amazonian biodiversity. If pre-Columbian urban settlements that formed ADE were as large and as densely occupied as assumed in this volume, there was little tropical forest left within and around the sites. Ethnobotanical analysis demonstrates that there was little or no forest present at the ADE of Araracuara when occupied and farmed (Mora et al., 1991).

The vegetation within and around the large ADE described in this volume was probably field and managed forests of tree crops. "Natural" biodiversity may have been completely replaced by agro-diversity on these ADE (borrowing from Brookfield, 2001). Historical ecologists and New Ecologists question the received wisdom that old forests with continuous canopy, the traditional image of ideal wilderness and optimal forest succession, are necessarily the only or best kind of diversity or the healthiest environment (e.g. Botkin, 1990; Zimmerer and Young, 1998). Graham (1998, 1999) argues that high biodiversity in the Maya region may be due to pre-Columbian urbanism rather than its absence.

Although unproven, ADE probably supports higher biodiversity and biomass, especially after abandonment of the site for settlement, than non-ADE landscapes.18 Various authors (Clement et al., 2003; German, 2003: Major et al., 2003; Mora, 2003) stress that it is possible to measure modern biodiversity on ADE. Inferring pre-Columbian biodiversity and agrodiversity in Amazônia is more difficult but possible (e.g. Mora et al., 1991; Piperno and Pearsall, 1998; Stahl, 2002). As authors of this volume have shown, soil biota and beneficial biological processes are enhanced in ADE and the diversity of soil organisms should be included in calculations of ADE biodiversity.

AMAZONIAN DARK EARTHS AND CONTEMPORARY SOCIETY

Pre-Columbian urban societies in the-Amazon figured out successful and efficient ways to sustain large and dense populations, produce surpluses, and manage wastes.

Is ADE relevant to waste management and agricultural development in contemporary Amazônia? Two questions must be asked: "How did they do it?" and "Can we replicate the process?" The authors of this volume have clearly demonstrated that ADE has much greater agricultural potential than typical non-anthropogenic soils in the Amazon Basin. The possibility of supporting current and future populations in Amazônia on smaller units of land through agricultural intensification is attractive, especially when framed in terms of conservation and management of biodiversity and cultural diversity. The archaeological record shows that ADE sites were occupied by large populations for long periods of time in what are often referred to as marginal environments, a clear indication of success and sustainability. Thus, deciphering the "secret" of ADE formation and maintenance may have considerable modern applied potential.

The huge and rapidly expanding "jungle cities" in the Brazilian, Ecuadorian, Colombian, Bolivian, and Peruvian Amazon produce prodigious quantities of organic urban waste. In the recent past, garbage was simply disposed of by tossing it into the rivers, lakes, or around settlements. Today, disposal and management of garbage has become an important aesthetic, health, and development issue. The systematic recycling and use of the organic matter component of urban garbage is relatively new. Tons of Brazil nut shells produced by the Brazil nut extractive industries are used for generating electricity to supply the city of Riberalta, Bolivia. Urban organic wastes have been used to restore the dry tropical forests of the Guanacaste region of Costa Rica (Janzen, 2002). Authors of this volume (Hiraoka et al., 2003) provide additional examples of management.

ADE is an excellent example of the capture, processing, composting, and recycling of organic and inorganic soil nutrients on a huge scale in pre-Columbian urban contexts. What can be learned from the past and what of this knowledge can be applied today? Several authors promote the ideas of using urban garbage to create new productive ADE. The pros and cons of applied ADE research are briefly addressed in this volume. Looking to the past (in this case, applying pre-Columbian technology) to resolve today's problems is often criticized as naïve or worse. The historical processes, technologies, and environmental, demographic, and cultural contexts that produced ADE may be completely alien to the contemporary world. The production, management, and use of ADE raise some obvious health, economic, and aesthetic issues.

The wastes produced by urban populations in Amazônia today are loaded with toxins, synthetics, and diseases; quite different from those produced by native peoples. Sombroek (1966: 261) and Sombroek et al. (2003) discuss the economic issues that need to be addressed to reestablish or create new ADE. Sombroek et al. (2003) conclude that it is theoretically possible but probably not economically feasible due to high labor, storage, and transportation costs relative to benefits. What can be done to encourage rural and urban peoples to participate in ADE studies and applications? Certain incentives such as those suggested for carbon sequestration by international agencies could be used to promote ADE. According to Sombroek et al. (2003), an increase of 5-10% of ADE would mean a significant sequestration of atmospheric carbon in the soil. According to the authors in this volume, an important process in the creation of ADE was massive and sustained incomplete burning to produce charcoal that was incorporated into the soils. Lehmann et al. (2002) promote "slash and char" as a means to this end. The conversion of urban organic wastes into useful charcoal is intriguing, but has obvious disadvantages (Madari et al., 2003). Large-scale burning is considered by nearly everyone (environmentalists, conservationists, green politicians, and development planners) to be the anathema of contemporary environmental resource and biodiversity management. Charcoal production throughout the world is blamed for much atmospheric degradation and deforestation.

FUTURE DIRECTIONS AND POTENTIAL APPLICATION

As pointed out by many authors of this volume, most claims about long-term fertility, agricultural potential, and sustainability of ADE are anecdotal. Well-designed experiments to reproduce ADE based on insights from archaeology, geography, ethnography, history, agronomy, soil science and other disciplines as discussed in this volume are critical. Long-term experimental ADE plots and controls should to be established in native communities and agricultural stations. Experiments such as those reported in this volume are a step in the right direction (Lehmann et al., 2003a; Sombroek et al., 2003). Intensive agriculture such as permanent orchards and house gardens may provide better models for ADE formation, use and maintenance than slash-and-burn (Denevan, 2001; Hecht, 2003; Hiraoka et al., 2003; Myers et al., 2003).

We propose the following issues as research priorities:

(1) *Define Criteria for Identification of ADE and its Variants:* A widely acceptable suite of archaeological, physical, and chemical criteria and inexpensive efficient methods for the identification of ADE are needed. Also, soil criteria will permit accurate identification and mapping of terra mulata or Preceramic period sites where potsherds may not be present. According to many authors in this volume, terra mulata are probably more common than the classic terra preta soils.

(2) *Systematic Archaeological and Soil Surveys of the Amazon Region:* Despite the recent excellent research on ADE, we still do not know the total geographical extent or full range of variation of ADE. Systematic archaeological and soil surveys to create a basic inventory of ADE in a Geographic Information System (GIS) would be a good start. GIS would also be an excellent tool combined with multispectral analysis of aerial and satellite imagery to efficiently map ADE. Predictive modeling could be done remotely once the unique signatures or indicator species of the vegetation canopy on ADE are defined (e.g. Balée, 1989; Clement et al., 2003, German, 2003, Major et al., 2003).

(3) *Cultural Resource Management Plan for ADE:* As archaeological sites, ADE should be identified, registered, mapped and evaluated as cultural resources. The systematic surveys reported in this volume are a good beginning. Management implies careful assessment of the impact of

urban and agricultural development on ADE and steps that can be taken to conserve them for the future.

(4) *ADE Content Analysis:*-Although many of the elements that make up ADE have been documented through archaeological, ethnobotanical, and soil analyses, the total range of components and their frequency remain elusive. In addition the contribution of different sources of organic material need to be quantified. Sophisticated analysis of pre-Columbian biodiversity from middens (Stahl, 2000; Mora, 2003) and fine-grained analyses in this volume provide examples of how this could be done.

(5) *Ethnoarchaeology, Ethnography, and Ethnoscience of ADE:* Although most contemporary scholars do not assume that the present-day ethnographic record represents the past, studies of native peoples and rural farmers can provide important analogies or models to test against the archaeological record (David and Kramer, 2001). Indigenous knowledge about fauna, flora, and climate is better known than soils in the literature on Amazônia. Studies of indigenous and modern folk classifications of soil, life history of artifacts, cultural views about garbage and sanitation, patterns of disposal, midden formation, and recycling in traditional societies would augment those presented in this volume.

(6) *Experiments in ADE Formation, Maintenance, and Use:* Archaeologists have learned that experiments are often the only means of understanding past technologies and indigenous knowledge systems, especially those that have no modern analoga. Experiments in construction and use of abandoned agricultural systems such as raised fields and terraces have provided important insights. The controlled experiments proposed by various authors in this volume will be valuable.

(7) *Comparative Research on ADE:* Is ADE a unique Amazonian phenomenon or is it similar to black earths reported in other parts of the world? Cross-cultural analogy regarding ADE and other anthropogenic soils in a global perspective remains under investigated. Woods (2003) briefly discusses

ADE in a comparative perspective. The vast literature on anthropogenic transformation of the environment may provide insights (e.g. Fairhead and Leach, 1996; Redman, 1999; Brookfield, 2001).

(8) *Computer Modeling and Simulation of ADE:* Many of the unanswered questions about ADE may be answered through modeling of carbon sequestration and carbon cycling as proposed by Sombroek et al. (2003). Multiscale modeling from local to regional to global scale may be possible. Because ADE and other strategies used by pre-Columbian Amazonian peoples impacted most of the basin, robust and powerful models will be needed. The experiments, ethnoarchaeology, and mapping of ADE could provide the basic data to construct and test these models.

(9) *Biodiversity and ADE:* Countering contemporary ideas that most human interaction with the environment are negative, several authors of this volume point out that ADE may have actually improved biodiversity in both the short and long term. More detailed comparative studies of ADE and adjacent non-ADE could test the premise proposed by historical ecologists that human activities can increase biodiversity. The hypotheses posed by Clement et al. (2003) regarding ADE as sites of agrodiversity and genetic reserves of useful plants need to be investigated.

CONSERVATION, MANAGEMENT, AND VALUE OF AMAZONIAN DARK EARTHS AS CULTURAL HERITAGE

Most obvious contemporary uses of ADE are for farming and as mines for construction fill, potting soil, and archaeological treasures (Woods and McCann, 1999; Hiroaka et al., 2003), both of which are incompatible with the importance of ADE to the scholarly community as archaeological sites and cultural heritage, stores of indigenous knowledge, locations of biodiversity, and carbon sinks (as discussed in this volume). As an archaeologist, I am concerned about the continuing destruction of ADE as cultural resources, a view shared by many of the authors of chapters of this volume. The looting of artifacts in ADE for private collections and museums is particularly disturbing. The report of over 300 urn burials destroyed at a single ADE site near Manaus is sobering (Clement et al., 2003).

The distribution of ADE unfortunately often corresponds to areas of contemporary urbanism, road building, agricultural expansion, and population growth especially in the Brazilian Amazon. The Taperinha site near Santarém with the earliest pottery in the Americas and the oldest documented ADE (7,000 BP) has been removed as construction fill (Roosevelt et al., 1991). Hiraoka et al. (2003) and Woods and McCann (1999) document the mining of ADE for commercial soils for urban gardeners. It is possible that when the insights of the scholars (this book for example) are more widely publicized, the exploitation and destruction of ADE will increase. As most of the authors of this volume would agree, ADE under poor management is not a renewable or sustainable resource. A proactive stance by scholars and the public is needed because the process of destruction has already begun in earnest.

Can existing ADE be sustainabily used and managed? Could new ADE be created to take the pressure off the original ADE? Do representative ADE deserve protection and management as cultural heritage? As archaeological sites and national cultural patrimony, ADE ideally fall under national and international legislation designed to protect, manage, and interpret cultural resources. Cultural

Resource Management is new to the Amazonian region and difficult to operationalize given poor funding and inadequate promotion of its benefits. Local governments, farmers, and landowners would probably reject taking valuable ADE farmland out of production for its protection. Because local and national protection is unlikely, could ADE fall under the protection and management of international agencies? The World Heritage designation of archaeological sites, monuments and landscapes by UNESCO is intended to protect and manage scarce cultural resources. To be nominated as a World Heritage site or cultural landscape, the location must be "unique" and of "outstanding universal value" (UNESCO, 2002). According to UNESCO's operational guidelines, ADE could fall under the category of cultural landscape or archaeological site. UNESCO designation provides funds and training for local protection and management of nominated locations.

- Are pre-Columbian ADE, modern settlement, and farming necessarily incompatible? Lived-on and farmed ADE may be the best means to protect them. The simplest solution may be the continued occupation and use of ADE by native peoples and small farmers for settlement and farming which may have a positive effect in protecting, maintaining and regeneration of ADE.

One of the largest ADE is possibly preserved under the cities of Santarém and Belém (Nimuendajú, 1952; Smith, 1980; Roosevelt, 1991, 1999; Kern et al., 2003). Promoting, interpreting, and protecting ADE as cultural heritage to the public, government, and international agencies will be a challenge. Some of the strategies for promoting eco-and cultural tourism could be adapted for protection and management of ADE. Triple alliances between archaeology, native peoples, and conservation of biodiversity have been forged with some success in Latin America (Kayapó, Sirionó and Xinguano examples). The potential links between biodiversity and ADE could be exploited for protection under traditional nature parks and reserves. Clement et al. (2003) suggest that the diverse species of economic plants, feral domesticates, and crops growing on relatively undisturbed ADE may be "genetic reservoirs" of landraces thought lost with depopulation and European conquest. We can assume that the biodiversity and agrodiversity where ADE is regularly mined and farmed for market crops has probably been lost forever.

A less tangible, but important "value" of ADE for contemporary society is as a record of an indigenous knowledge system, an ethnoscience, or appropriate technology, in this case a possibly sustainable landuse strategy that has become lost over time. Publications about ADE by Charles Mann (2000a, 2000b, 2002a, 2002b) and the documentary Search for Eldorado (DOX Productions, 2002) are examples of how the popular media can be used to capture the public's imagination. The idea that native peoples constructed agricultural soils where none existed is a powerful concept. As Mann points out, the Woods and McCann (1999) hypothesis about living and reproducing anthrosols has considerable appeal with both the public and scholars.

Other potential "values" of ADE research are the protection, promotion and regeneration of the native culture of the descendants of the people who made ADE. Documentation of indigenous resource management (Ka'apor; Kayapó examples) increases the appreciation of indigenous knowledge, countering the idea that native practices are backward, primitive holdovers from the past. On the other hand, over-promotion can result in creating new Myths of the Noble Savage.

CONCLUSIONS

As stressed in this volume, ADE should not be treated as simply a soil science phenomenon. It is obvious that the formation and maintenance of ADE are the result of a complex and dynamic interaction of culture, settlement, technology, and production strategies. Continued multidisciplinary, interdisciplinary research as exemplified in this volume, is the best approach to understand ADE. These studies combine fieldwork,participant observation, and bench science of archaeologists, anthropologists,geographers, scientists, geneticists, and botanists. The authors of this volume stress the importance of the human dimension in our understanding of soil formation, use, and management. Amazonian peoples took soils that are generally considered marginal in terms of nutrient availability, texture, drainage, and depth and turned them into productive farmland through their management of organic matter from urban settlements.

Increasing knowledge of ADE has important implications for the interpretation and understanding of Amazonian prehistory and history. In contrast to traditional views stressing environmental limitations for cultural development, ADE is evidence of large dense populations, permanent residence, and sustainable lifeways that continue to effect contemporary inhabitants of the Amazon. ADE is one of many strategies developed by Native Amazonians to thrive in what has been called a counterfeit paradise.

ADE was one of many strategies used by native peoples to transform the Amazon Basin. Historical ecologists argue that all environments in Amazônia are to some degree anthropogenic and have complex, dynamic human histories. These transformations probably go back to the first human colonizers of the region over 10,000 years ago who brought with them one of the most powerful pre-industrial tools for landscape management: fire.

Foragers, settled farmers, and later urbanized town inhabitants from the late Pleistocene to the present have played a role in shaping the Amazonian landscape as we know it. ADE are just one of many anthropogenic soil modifications which included creation of physical infrastructure such as terracing, raised fields and more subtle regional transformations such as induced upslope erosion and downstream sediment capture, fish weirs, earthworks for settlements, transportation, and agriculture (Balée, 1994; Stahl, 1996; Erickson, 2000a; Denevan, 2001). Native American manipulation of the genetics, number, and distribution of crops and non-domesticated economic species; house gardens and fields; and agroforestry are examples of indirect modifications of soils and local environments.

4

Early Human Settlements as an Opportunity for Infectious Microorganisms

INTRODUCTION

Emergence of new infectious diseases has been observed and documented in the 20th century, but the phenomenon must have occurred throughout history. The study of the factors that may be involved in this process is in its early stages (May et al 2001; Woolhouse and Dye 2001). Nevertheless, it is worth thinking of the possible implications of new theories in this young field to our understanding of historical events. Textbooks in history of medicine tend to highlight the importance of the emergence of infectious disease in humans as coinciding with the formation of settlements in the historical era (Fiennes 1978; Kiple 1993; Crosby 1994). It is believed that as humans gradually formed larger groups, settled down, and assumed more control over the growth of foodgrains, they also began to domesticate some animals. Thus they were exposed to an increased risk of acquiring animal-based infections (zoonotic infections).

Close and persistent contacts between humans and domesticated animals, as well as changes in human diet, are generally considered central to the acquisition of new infections and diseases. This line of explanation is focused on humans, and the impact of the new ecology on animals and infective microorganisms tends to be left unexplored. I wish to argue in favour of a conceptual shift from the traditional point of view to one that is less anthropocentric and is focused more on the impact of human settlements on the infectious agents. This approach is justified because it places both humans and microorganisms in a much more encompassing natural environment and allows us to consider how both may have been subjected to natural selection during the period of early domestication. This approach is also supported by recent theories of the emergence of infectious disease.

INFECTIOUS AGENTS AND HUMANS

The arguments presented in this paper are based on characteristics of infective microorganisms and humans and the interaction between them. Excellent discussions of several aspects of this interaction as well as historical and current examples are available (Dobson and Foufopoulos 2001; Baum and Bar-Gal 2003; Ewald 2003). Most infectious microorganisms are thought to be host-specific or limited to a few hosts. The restriction of an infection to a certain host is part of the evolutionary adaptation process that the host and the infective agent have undergone over a long period of time. A change in the 'habitat' for an infectious agent implies changes in the host or the infectious agent, or the introduction of a third party facilitator (e.g. a phage). A change in the host that is not detrimental to that host, that is, a permissive change, may take a huge number of generations. Given the vastly longer generation time of humans compared to microorganisms, it is more likely that such a change would occur in the infective component rather then in it human, or animal host.

THE HISTORICAL SETTING

Changing conditions resulting from early human settlements would have impacted those involved in that ecology-humans, domesticated animals and microorganisms. In the pampas or prairies of our ancestors, the most likely contact between species was on the feeding chain. Humans and animals had nomadic life styles, thereby minimising the chances of cross contamination. Apes or hunter-gatherer humanoids are likely to have harboured about the same number of infections as any other large migratory animal (Kiple 1993; Martin 2003). But each species would have acquired its own infectious agents and diseases; an immune system with a different vocabulary and skills, as it were. The exposure of members of a species to a new infection could have had a fatal outcome as happened in the case of the massive epidemic that followed the exposure of native North Americans to infections that were, up to that time, limited to the European population (Kiple 1993).

As humans domesticated animals and became settled, they continued to improve their ability to control the growth of food in their immediate surrounding and maintain a supply of food throughout the year. This persistent, close contact of humans and animals gave rise to a new situation: infectious agents could be transferred and shared among animals of different species and humans. This period constituted a crucial step in the development of human societies, and it was accompanied by a noticeable alteration in the diseases that affect humans (Kiple 1993). Nevertheless, the relevant ecological changes would have been of more significance for the infective agents rather than for the humans. With their rapid generation turnover, the infectious agents were much better equipped to take advantage of this new ecology by using animals and humans as their explorative 'laboratory'.

Studies of ancient infections are limited by the degradation of biological material, but recent technologies promise some insights (Cohen and Crane-Kramer 2003; Rothschild 2003). Nevertheless, the possible outcome of this historical change in human habitat remains speculative. It is likely that in this process, and over time, there have been infectious agents that successfully adapted and moved from humans to animals and vice versa.

There must have been infectious agents that successfully moved from one animal to another and, later, with further modification, 'jumped' to humans. Some domesticated or partially domesticated animals must have come in contact with their wild relatives, thus allowing for the transmission of newly modified infectious agents from the human habitat to the virgin wild habitats in forests or savannahs. In parallel to known human history (epidemics following the European explorations in the New World), the re-introduction of such modified infectious agents to the wild habitat may have had disastrous consequences to infected animals. Furthermore, some of the animal infections that were contracted originally from humans could act as reservoirs in the wild that could come back later and infect future human generations.

RECENT PERSPECTIVES

Some of the claims made in the previous section can be supported by observations. Rapid changes in microbial or viral flora in response to antibiotic therapy are well documented in current medical practice. Familiar examples include changes that result from the evolution of drug resistance. Such changes are usually facilitated via the acquisition (importation) of foreign genetic material from other infectious agents in their immediate environment. The likelihood of this happening is enhanced if the antibiotic fails to eradicate the whole stock of bacteria, or when there are many bacteria in the environment that already carry the resistant factor. The former situation corresponds to sub-optimal therapy; the latter is often seen in hospitals where, via plasmid exchange, generations of bacteria come to possess the information of which antibiotics are present in the environment.

Another example comes from bacteria that can alter their life cycle and form a more resistant stage under hostile environmental conditions (physical, chemical, or biological); for example, L-form or cell wall deficient bacteria (Domingue and Woody 1997). The relation of these altered forms to disease is not well understood but it has been speculated that the alteration may result in altered disease processes and manifestations (Domingue and Woody 1997). We do not know what, if any, the consequences of such processes may have been for human history.

"[Most] ecological literature has paid little attention to invading microorganisms, including viral, bacterial, and protozoan agents of infectious disease. Studies of 'emerging disease' are relatively et al 2001). The studies alluded to emphasize that pathogens can cross the species barrier, but this does not necessarily imply a successful establishment of an ongoing infection. Many factors influence the probability of success. They involve, among other things, the presence of other infections in the host (cross-immunity), the size of the susceptible population (measles virus requires a population over 300,000 to avoid fading out), and the availability of an alternative reservoir for the microorganism (May et al 2001).

Out of 1415 species of microorganisms pathogenic to humans, 868 (61%) are zoonotic infections. Interestingly, the majority (75%) of emerging diseases are of zoonotic origin (Taylor et al 2001).

While direct transmission from animals is important for some zoonotic pathogens, the majority of humans are infected by other humans (Taylor et al 2001). Multi-host pathogens are prevalent among human pathogens and even more so among domestic mammal pathogens (Cleaveland et al 2001). An understanding of the dynamics of infectious disease in complex multi-host communities may help to mitigate the threat of infectious diseases to both humans and animals.

Current molecular biology technologies allow scientists to reconstruct the evolution of micro-organisms. The reconstruction process offers challenges and remains somewhat speculative but can yield interesting insights (Simmonds 2001; Rothschild 2003). Several examples are now available for the reconstruction of the evolution of microorganisms over time, their migration through different species and over different geographical areas. These include recent emerging diseases such as HIV and hepatitis as well as tuberculosis and syphilis, diseases that had their emergence much earlier in our past.

SUMMING UP

The purpose of this commentary was to expand the traditional anthropocentric view of human infections and consider the possible implications of such a conceptual shift. I have tried to point out that the settlement of humans (and domestication of animals) could have resulted in the emergence of new diseases of animal populations. The diseases could then have spread from the domesticated animals to the 'wild' animal reservoir. This line of argument reflects the fact that human history is an integral and interdependent component of natural history.

5

Environmental Factors in the Chronology of Human Evolution and Dispersal

INTRODUCTION

This is a four-year mini-thematic programme aiming to integrate and strengthen UK research in human evolution. The rationale underlying the proposed research and the timing of this programme is the acute awareness that studies of human evolution need to be tied closely to the ecological and environmental expertise underpinning Quaternary science generally, and that in particular research among the disciplines of palaeoanthropology, archaeology and palaeoecology needs to be integrated in an effective manner to realise these wider aims. This programme will be highly interdisciplinary and has two parallel pathways. Firstly it will acquire and analyse new field data, thereby making a major contribution to human evolution and quaternary science. Secondly it will exploit existing datasets from archaeology, palaeoanthropology, palaeoecology, as well as genetics, within a developing E-Science context.

The pathways lead to several key issues relating to a fundamental, curiosity driven question:

" What impact did environmental factors have on human evolution and dispersal?

EFCHED aims to explore the interface between the disciplines engaged in human evolution research. These include the classic ones of human evolutionary studies-palaeoanthropology, archaeology and palaeoecology-as well as emerging ones such as evolutionary genetics and palaeoclimatic modelling, as well as allied disciplines such as neoecology (Figure 1). The relationship between environment and evolutionary change will be at the heart of EFCHED projects. Proposals will be judged by their commitment to the role of interdisciplinary approaches in investigating such fundamental relationships.

The mini-thematic programme will serve as a catalyst for a new integrative science within the field of human evolutionary research, and will provide training opportunities for the interdisciplinary postgraduate and postdoctoral researchers that the field urgently requires. The programme's output will be of wide public interest and serve as a showcase for NERC science policy and process.

BACKGROUND

The background for this programme is:

- The UK has an internationally-recognised group of researchers working in this field. Our expertise is particularly strong in human morphology and palaeontology, Palaeolithic archaeology, Quaternary faunas, science-based dating, ancient biomolecules and human genomics. In all these fields we have led the way in developing the environmentally-based approaches that are now becoming a priority.
- The growth in evidence related to both human evolution and its environmental context leading to a need for more large-scale projects that can pull together case studies.
- The growth of molecular approaches to evolution which have opened up new issues (such as the debate about hominid and human origins) which depend upon timing and environmental context for resolution.
- The development of major new techniques, especially relating to environmental context and reconstruction, that have greatly expanded the data available and approaches possible to human evolution.
- Recognition that the relationship between humans and their environment needs consideration of

organismenvironment interactions at both ecological and evolutionary time scales.

- The great public interest in human evolution, evidenced in major TV documentaries, books and newspaper articles.
- The opportunity to exploit the growing availability of major palaeoanthropological and archaeological databases, programmes such as Rapid Climate Change (RAPID), and to bring a greater analytical dimension to studies of human evolution.
- The growing interactions between archaeologists, palaeoanthropologists, palaeoclimatologists and molecular biologists and the need to provide a new disciplinary framework in which they can work together.

NEED FOR A MINI-THEMATIC PROGRAMME

The EFCHED mini-thematic programme has a number of advantages over non-thematic funding:

- Providing a focused scientific response to the challenges posed by recent developments in human evolutionary studies, all of which have a strong environmental component.
- Providing the framework for strengthening long-term research on human evolution.
- Integrating the three essential components of human evolutionary research-primary field projects generating original evidence, laboratory analysis, and synthesis.
- Providing a sharper focus on evaluating and addressing key evolutionary questions.
- Providing a formal link between two fields that have grown together over the years-i.e. human-centred evolutionary and archaeological studies and quaternary science.
- Allowing greater exploitation of emerging scientific techniques, especially science-based dating, ancient biomolecules, evolutionary genetics, and palaeoenvironmental/climatic reconstruction.
- Providing a better way of integrating the development of scientific techniques (one of the great strengths of SBA) within problem-driven research.

Recent Achievements in Human Evolutionary Studies

Listed below are 10 examples of key discoveries and developments in human evolutionary studies that prompt environmentally-based approaches:

1. The recognition of more than 15 species in the hominid fossil record, leading to questions about the environmental and climatic factors leading to speciation, extinction and isolation.
2. The fact that bipedalism and brain enlargement evolved independently, at widely separated times, and in assorted environmental conditions.
3. The identification of a major shift in body size and shape with the evolution of Homo, compared to Australopithecus, with the possibility that this is related to a major shift in the way the environment was used.
4. The discovery through isotope analysis that the australopithecine diet may have been far more varied than had been inferred from classic morphological studies.
5. The realisation that humans first occupied almost three quarters of the earth during only the last one per cent of the time since hominids first appeared
6. The recognition that stone tools remain remarkably stable over long time-spans, and that technological solutions to survival among earlier hominids were not habitat specific. This implies that human culture is contingent upon both environment and phylogeny.
7. The finding that there was a significant level of geographical isolation and evolutionary divergence among Middle Pleistocene hominids, and that this may have been related to changing environmental patterns associated with the onset of glacial cycles.

   EFCHED: proposal for a NERC thematic programme 07/02/2003 3
8. The recognition, derived largely from molecular genetics that hominid evolution may have been characterised by extreme demographic instability, and that climatically driven population contractions and expansions may have given rise to our own species.

9. The proposal that the early divergence of modern human populations may have been influenced by volcanic as well as climatic events.
10. The discovery, through both archaeology, fossils and genetics, that the origins of modern humans is a localised African event, closely connected to the last two glacial cycles.

RESEARCH AIMS AND METHODS

The aim of the proposed theme is to encourage research that formulates and tests hypotheses on several chronological and spatial scales that bear on the causal relations between environmental and climatic change and human evolutionary patterns-genetic, morphological, behavioural and cultural. These relationships are epitomised in the questions posed at the outset, and from which more specific hypotheses can be derived-i.e. 'what impact did environmental factors have on human evolution and dispersal?'.

The hypotheses will be tested through interdisciplinary projects that will include:

- The compilation of relevant archives and further enhancement via fieldwork, collections analysis and experiment.
- The analysis of evidence at a variety of geographical and temporal scales using a number of approaches including modelling, and comparative and experimental studies.
- New fieldwork to generate new data to answer specific questions
- Development of new syntheses based on comparison across time scales and regions.

RESEARCH THEMES

We identify two major related themes as lying at the heart of the EFCHED programme: " Humans in an evolutionary context " Humans and dispersals. Each of these sub-themes poses research problems that involve archaeologists, palaeoanthropologists, palaeoecologists, palaeontologists, molecular geneticists, as well as earth and life scientists more generally, although for each of them the mix and emphasis varies. These themes drive the selection of areas and time periods which will deliver appropriate data sets to achieve the programme aim.

Humans in an Evolutionary Context

It is second nature to archaeologists and palaeoanthropologists to be aware of the long term and dynamic nature of the human lineage, but this is still something that has not been operationalised through into many dis ciplines. Looking at humans in an evolutionary context-i.e. as a series of different species, with different morphologies, behaviours and ecological contexts-and using the excellent archaeological and fossil record documenting hominid presence (and absence) over a period of nearly 5 Myr allows a number of issues to be explored:

- Macro and microevolution: some hominid species living more than 4 million years ago were very ape-like, and their record is no different from that of any other mammalian Cenozoic lineage; others, such as Neanderthals, are richly recorded in both tools and bones, and have a span of only 200,000 or less. The process of transformation from one sort of creature to the other (when and how did it occur?) will be a challenge to many theoreticians and empirical scientists.
- The wealth of the hominid fossil record permits issues such as the timings of speciation and extinctions to be examined.
- The fossil record records the history of human morphological adaptations, and this can be used both to examine such major evolutionary issues as rates of change and also the actual nature of human biological adaptation.
- The archaeological record (e.g. stone tools) provides the potential to study lineage persistence and distribution, and behavioural adaptation, not found in other species. As a result it potentially provides one of the best test-beds for general evolutionary hypotheses.
- A critically important and rapidly emerging field is that of evolutionary molecular genetics, and this is being used to reconstruct the phylogeny of humans and other species,

to estimate coalescence events, and reconstruct prehistoric demographic patterns. However, there is considerable controversy about the underlying chronological assumptions, based on mutation rates and other internal inferences. The Pliocene and Pleistocene fossil records can be used to provide a broad comparative dataset through which calibration of molecular systems can be tackled.

- The development of bioinformatics has led to the potential to develop more sophisticated and analytical uses of chronological data (e.g. radiocarbon lists, etc.). This database approach ('palaeoinformatics') can be used to examine in detail the chronological patterns of human distributions in the environment.
- The period in question (5-0 Myr) covers times in which very different chronological resolution is possible, from the early part where chronology is measured in 105y to the recent end, where a resolution of 102y may be possible. Within these scales is the fact that some archaeological context s allow 15 minute activity episodes to be observed. One of the tasks for an evolutionary explanation lies in the integration of different time scales.
- The growth of palaeoinformatics allows testing across dating methods and scales, and thus develops a more stable chronological framework for scientists from a number of disciplines.
- The impact of rapid climatic change on human society, as revealed for example in the chronological precision now provided by the Greenland ice cores, can now be investigated at a variety of scales and the outcomes observed.

HUMANS AND DISPERSALS

The course of human evolution sees hominids changing from a group of geographically restricted African species to a single global species. There is thus a strong geographical, and therefore ecological, element to hominid evolution, and a key aim will be the investigation of the patterns and processes by which hominids repeatedly dispersed and contracted across landscapes at all scales.

A major question to ask is, what role did environmental change play in the observed pattern? Understanding the geological context at a variety of scales and processes provides the critical framework for this theme

- Dispersals are strongly influenced by environmental effects, and it will be a major part of the theme to investigate the timings and distributions of dispersals in relation to local and global patterns of climatic/environmental change.
- Dispersals are also strongly influenced by the behaviour of the hominids, including their technology, and comparing patterns of dispersals across different human populations will show the role of changing behaviour in the distributions and persistence of human groups.
- Dispersals and contractions form the basis for biogeographical patterns, for both humans and other species. Allopatry, isolation, and admixture are all major evolutionary factors, and identifying corridors of dispersals, refugia, and contact zones will be of use to ecologists, geneticists and other scientists.
- As populations disperse-or contract into refugia-they come under new selective influences and contexts for genetic drift. Hominid morphology thus has a strong geographical component that can only be understood by bringing together ecological factors with the historical contingencies of the populations.
- At the macro-level it has been proposed that pulses of speciation and extinction have been directly related to major climatic change. The theme will explore this issue in terms of geographical variation, taxonomic differences, and rates of change.
- Embedded within global patterns of change are local processes, varying from regional climatic patterns, shifts in sea and lake levels, and tectonic/volcanic effects. These are the ones that directly impinge on human and other populations, and one focus will be the discerning of such local patterns within larger scale changes.
- During most of human evolution the direction of influence has been the environment acting as the selective pressure

shaping human morphology and behaviour. The morphological and behavioural patterns observed in the fossil record will be screened for the effects of environmental change.

- With the emergence of modern humans and the global expansion of population the direction of causality shifted, and humans increasingly influenced their environment. Mapping this process and its effects within a geological framework will be a major area of research

DELIVERABLES

The principal deliverables relevant to the wider scientific and public community will be:

- A new understanding of our position as a dominant global species and implications for our future.
- An improved understanding of rates and scales of climate and landscape change, and how these impacted both on human evolution and dispersal.
- A more precise appreciation of how human evolution is affected by rates and scale of environmental change
- An evolutionary context in which the results of post-genomic research will be understood.
- Stronger co-operation between collaborating disciplines engaged in human evolution research. The principal benefits for human evolutionary studies and SBA will include:
- Achieving greater confidence in the correlations between the archives of archaeological, palaeoanthropological, genetic, palaeoclimatic and palaeoenvironmental data.
- Improved knowledge of the adaptive and phylogenetic diversity of hominids and their cultures set in a palaeoenvironmental context.
- A more detailed understanding of the role of differences of scale, both temporal and geographical, in evolutionary development.
- Understanding the Quaternary legacy for modern human diversity, both morphological and genetic.

- A better-focused integration of the environmental factor in human evolution and the ways in which specific variables of landscape structure and dynamics have selected for, or been selected by, changes in behaviour.
- The development and testing of predictive models (using GIS) of the spatial and temporal distribution of hominid populations and their inter-connections at the global scale, using genetic, morphological and archaeological data as the control.
- Combining better population genetic models with information from independent and multiple sources of data to increase the value of human evolution as a framework for new genomic studies.
- Enhancement of modelling techniques to simulate how genetic and demographic structures can emerge from hominid palaeontological and archaeological empirical observations.
- Database of high-resolution terrestrial palaeoenvironmental records for key periods by integrating existing and stimulating the collection of new proxy records.

RELEVANCE TO NERC: NATIONAL AND INTERNATIONAL DIMENSIONS

NERC's Strategy for Science-Based Archaeology identifies three key initiatives which are directly addressed in this programme; Refining the chronology of the human past, Human evolution, ecology and behaviour and biomolecular archaeology. The proposal fits squarely into two of the six "science opportunities" identified by the Science and Technology Board sectoral strategies which are: Climatic and Environmental Change and Evolution of the Biosphere and the global environmental change area of core science. The programme builds on the previous SBASG Thematic Programme , the Ancient Biomolecules Initiative (ABI) and dovetails with the new RAPID programme. Recent NERC initiatives in E-Science will also impact on this programme.

Nationally, the Prehistoric Society in its 1999 report "Research Frameworks for the Palaeolithic and Mesolithic of Britain and Ireland", sponsored by English Heritage, identified the issues of dispersal as being central to future research. This report has been acted upon by English Heritage, which has commissioned further scoping studies for research investment. The Stage 3 Project, run through the Godwin Centre at Cambridge, provides a successful model for international, interdisciplinary research.

The international importance of human evolution is demonstrated by the National Science Foundation (USA), "Human origins initiative" currently under consideration as a ten year proposal. European interest in human evolution research is delivered, among others, through the ESF, CNRS (France) and the Max Planck Institute (Germany). UNESCO's World Heritage List contains sites relevant to human evolution in four continents. INQUA now has a sub-commission on climatic imperatives and historical contingency.

UTILITY AND BENEFICIARIES

A prime benefit of this Programme would be an increased and intensified interaction between diverse scientific disciplines encompassing a wide range of environmental, biological and historical sciences. Of necessity, this requires a world-wide perspective and offers exciting possibilities for international collaboration. It also demands explicit recognition that the collaboration of experts in fields outside archaeology and anthropology is more effective when the various investigations are formulated to include objectives that are of direct interest to the supporting sciences. The co-ordination through EFCHED of these potential co-funding partners will benefit NERC by extending the application of skills and the shared use of facilities and equipment. The proposal builds on recent NERC, JIF and Wellcome investment (the Radiocarbon Accelerator and Ancient Biomolecules Centre at Oxford) the AHRB Research Centre for the Evolutionary study of Material Culture at UCL and Southampton and Cambridge's £5m award from Wellcome, Leverhulme and SRIF to establish a Centre for Human Evolutionary Studies, all of which offer the potential for training PDRAs with an interdisciplinary focus.

Furthermore, the proposed programme will be of direct benefit to all scientists interested in human response to global climate change, and ultimately to policy makers. It will also be of direct relevance to organisations such as English Heritage, currently funding internationally important investigations of early hominid occupation on the Sussex coastal plain, and who are charged with managing and presenting the human heritage. EFCHED funded projects will be expected to produce materials which bring human evolution alive for the public. Involvement with the media to explain the process of science and the role of research councils such as NERC is seen as important to the programme.

TRAINING

The opportunity to train Postgraduate and Postdoctoral researchers within an interdisciplinary research programme is a key objective of EFCHED. The programme will break down disciplinary boundaries both through the requirements to train skilled practitioners in the structure of the varied data archives, and also in the range of scientific techniques and methods which are needed to interrogate them. The programme provides the potential to shape the next generation of researchers and equip them with the skills and interdisciplinary dialogues that only come through engagement with cutting-edge science. This will be achieved by tied studentships and project based PDRAs. It is to this cohort that we look for the new disciplinary framework focused on integrated human evolution rather than its current constituent parts. We predict that such a lead will also initiate a sea change in the structure of graduate training programmes. In particular we envisage inter-institutional collaboration that will share expertise and broaden the student experience. Human evolution is an international subject and we see opportunity through such initiatives as the World-Wide University Network, linking UK and US universities, to further demonstrate this claim.

RESOURCES

EFCHED is a two million four year mini-thematic programme starting in 2002.Within this budget there is a commitment for for data management of the bioinformatic archives. EFCHED will be organised through a scientific management committee which is drawn from all the major relevant disciplines. £120K is estimated to cover central project costs, which allow for a Town Meeting, archive standards conference, annual workshops, and web-site dissemination. The Chairman will act as programme co-ordinator and resources will be made available to support this role. The programme will be organised and resources allocated through one open call for project bids. The call will consider bids in the following areas: Bio informatics which assess, compile and calibrate archives, laboratory and experimental projects, modelling and fieldwork.

CO-FUNDING ARRANGEMENTS

AHRB has as a designated funding area the archaeology of human origins and through its Research Centres scheme offers further potential for co-funding. English Heritage's long term support for The English Rivers Palaeolithic Survey and the Boxgrove Project has been continued with projects, especially relevant to this programme at Cromer, the Sussex raised beaches and under consideration GIS based surveys of the Upper Palaeolithic and Mesolithic. The Wellcome Institute through its science based archaeology panel has a declared interest in human evolution. Leverhulme currently is calling for proposals relating to human evolution and archaeology. The BBSRC currently has a Bioinformatics programme and funding in this area is likely to expand across many funding bodies over the next few years. The British Academy has collaborative funding arrangements with CNRS that can be built on and a Larger Research Grant for funding pilot projects.

6

Human Ecology in Anthropology: Past, Present, and Prospects

INTRODUCTION

Anthropology is, by its very nature and tradition, a kind of multidisciplinary science. Sociocultural anthropology has history and humanities as components of its heritage, but is squarely situated within the social sciences.

Linguistic anthropology, in some ways, is even closer to the humanities than sociocultural anthropology, yet several branches of this field are quite empirical, and some are experimental, and can be allied with psychology. The archaeology branch of anthropology, while also firmly tied to social science, has somewhat closer linkages to the natural sciences for purposes of dating, faunal and floral analysis, and stratigraphy and geomorphology. Archaeologists also feel a close affinity to historians. Biological anthropology is often considered the least associated with social science, being linked to the biological, evolutionary, and biomedical sciences.

However, its real contribution to scientific inquiry comes from an integration of the biological and social sciences along with a long-standing interest in human evolution. These four branches of anthropology, despite a common concern with the central concept of culture and of social behavior, have quite different traditions of theory, training, method, and practice. The traditions of human ecology in anthropology tend also to be separated into their constituent subfields, although there have been efforts at integration in anthropology under the theoretical framework of human ecology. Some of these bases for division and integration of human ecology will be discussed.

CULTURAL ECOLOGY

Within anthropology, ecological approaches have been employed in a variety of ways. Cultural ecology has been applied in socio-cultural studies as an alternative to a deterministic application of "culture" as the primary causal agent leading to new "culture." In other words, culture, as ideas and behavior, can arise from the environmental circumstances (both social and physical) of people's lives or culture can arise sui generis, that is, in and of itself! Historical processes play an important role in this latter scenario. These two fundamental approaches to anthropological inquiry have characterized the science for many years. The functionalist school of anthropological theory, in which cultural attributes were identified as part of an interrelated system, was heavily criticized because it neglected historical explanation and human agency, that is human actions contributing to culture.

Environmental determinism and possibilism, functionalism, culture-area approaches, racism, evolutionism, and historicism were conceptual and theoretical perspectives that were all mixed in complex ways during the late 19th and early 20th centuries in anthropology. Franz Boas, the founder of American anthropology, demonstrated the influence of the environment on body size and form in migrants in the first decade of the 20th century, yet he rejected evolutionary explanations and identified human behavior and culture as arising from historical forces not environmental forces.

Early social studies of humans and their environment moved from the "environmental determinism" of the anthropogeographers (Ratzel, 1889-91; Semple, 1911), to the "environmental possibilism" of the ethnographers (Forde,1934; Evans-Pritchard, 1940), and to the "cultural ecology" of Julian Steward (1938, 1955). In the 1930s, Steward moved cultural ecology a step forward by rejecting the "...fruitless assumption that culture comes from culture..." (Steward, 955: 36). He also developed the concept of culture core as the behavior patterns most closely linked to the environment (e.g., subsistence and food acquisition). He advocated a three-fold analysis of relationships between (1) the environment and subsistence, (2) subsistence and behavior patterns, and (3) behavior patterns and other components of the culture, and his view of ecology was closely linked to the concept of "adaptation to the environment" (Vayda and Rappaport, 1968).

Later studies criticized Steward's "culture core" concept as too narrowly conceived. This form of criticism is quite characteristic of anthropology: rather than building on previous ideas and data, ideas are rejected sequentially as new theoretical approaches appear and rise in popularity.

Anthropological progress over the past century has been constrained because the pattern of exploration has been: first, limited application of scientific design and hypothesis testing; second, a continual succession of new theoretical frameworks and approaches without full exploration; and third, little validation of research results and limited development of a tested body of fundamental principles.

ECOLOGY WITHIN ARCHAEOLOGY

Within archaeology, interests in the environment date back to the 19th century. Eco-logical theory in archaeology tended to be linked to processes of culture change and evolution through the writings of Leslie White (1949) on unilinear evolution and Julian Steward (1955) on multilinear evolution. More recent interests in the 1960s and 1970s were in systems theory in behavior, and most recently, within the past decade or so, archaeologists have directed their interests toward regional climate change, historical ecology, and landscape ecology. It is also the case, as Karl Butzer (1990: 92) stated: that there has been an "..advantage of exposing archaeology to the intellectual cross-currents of anthropology. But it has also been disadvantageous, exposing archaeology to disciplinary fads and limiting effective contacts with other scientists."

ECOLOGY WITHIN BIOLOGICAL ANTHROPOLOGY

Biological anthropologists moved from 19th century and early 20th century typological approaches and race studies to the understanding of humans and their evolution via modern ideas about adaptation to the environment as a basis for understanding human variation. Ecological theory was tied also to evolutionary processes, but more in the realm of biobehavioral evolution and associated with the Darwinian concepts of "selection" and "adaptation to the environment" (Warren, 1951; Weiner, 1964). Ecological theory in biological anthropology became a fusion of evolutionary and ecological theory (Bates, 1953, 1960), along with ideas from environmental physiology (Dill et al., 1964), biogeography and human biogeography (Coon et al., 1950), demography (Spuhler, 1959), and human biology (Baker and Weiner, 1966; Lasker, 1969; Little, 1982; Little et al., 1990). Recently, socioecological theory has successfully been used to extend the synthesis between ecology and anthropology, focusing primarily on individuals rather than higher levels of organization (Winterhalder and Smith, 1981; Smith and Winterhalder, 1993).

Prominent in nearly all ecological theory in anthropology has been the concept of adaptation to the environment (Alland, 1975; Baker, 1966; Moran, 1979). Ecological studies in biological anthropology were stimulated in the 1960s by the work being done in ecosystems ecology by the scientists in the International Biological Program or IBP. At this time, there were several Human Adaptability Projects associated with the IBP that were influenced by systems science and efforts to modeling complex ecological systems.

TWO EARLY STUDIES

There were two very original anthropological studies that were done in the 1960s that were subject to a great deal of criticism in the anthropological literature. The first is the ecological study of the Tsembaga Maring of New Guinea by the late Roy Rappaport. The work was done in the central highlands of New Guinea. The second is the energy-flow study of Andean Quechua Indians of Peru by Brooke Thomas. This work was done on the Peuvian altiplano at a base elevation of 4,000 meters above sea level. In the first case, Rappaport was a single investigator with an overwhelming task that he set out for himself. This is a long-established tradition in anthropology for one investigator to live with a people and, through participant-observation, to learn about the workings of the society or culture.

Rappaport not only took on the job of describing and understanding the inner workings of Tsembaga culture, he also attempted to understand this in the context of Tsembaga ecology. In the second case, Thomas's task was no less daunting, but his work was done within the framework of an integrated project. This was the Andean Biocultural Studies project, initiated by Paul Baker (Baker and Little, 1976) at the Pennsylvania State University in the U.S., with the primary objective of studying the patterns of adaptation of high-altitude natives to the hypoxia and cold of the Andean altiplano. When Thomas began his work, several years of data on social conditions, nutrition, human physiology, demography, and weather conditions had already been collected, and the area had been mapped.

Rappaport's Work

Rappaport's research was reported in a now famous book entitled Pigs for the Ancestors, which was published in 1968, and reprinted in 1984 with a 190-page Epilogue, in which he addressed his critics and reevaluated some of the research (Rappaport ,1968, 1984). The work is brilliant, in that it addresses some of the fundamental issues underlying anthropological theory, including: social control, environmental causality for behavior, and the connection between individual behavior and cultural norms or prescribed social behavior.

In the work, Rappaport suggested first, that human population numbers, pig population numbers, the warfare cycle, agricultural productivity, patterns of exchange of goods, the distribution of land and people, and the maintenance of the ecosystem as a productive system were all tightly interrelated as a working system. Second, he suggested that the system was in a state of equilibrium maintained by feedback mechanisms. And third, and perhaps most controversial, that the regulating or controlling mechanism that kept the system going was the information provided in the form of ritual and a ritual cycle. Within this research, he took both a materialistic and a functionalist approach to social science, he identified human behavior as adaptive in the context of the social and ecological systems, and he identified human behaviors as subject to selection of favorable behaviors in the context of maintenance of the human/ecological system.

Needless to say, and despite the fact that ecological anthropology was in vogue at that time, Rappaport's critics in sociocultural anthropology were severe in their verbal assaults. His work had attacked some of the fundamental icons of anthropology. He and other ecological anthropologists were accused of:

(1) reifying the ecosystem (to treat the abstraction of an ecosystem as if it had material existence);
(2) vulgar materialism (a belief that the materialistic approaches used in ecological anthropology were simplistic in their social context);
(3) a calorific obsession (placing too much emphasis on flows of energy through the system);
(4) excluding historical factors (too much emphasis on equilibrium and stability in diachronic state in the systems studied);
(5) setting up false boundaries (human cultures go beyond ecosystem boundaries);
(6) shifting levels of analysis (applying one level of interpretation to another); and
(7) dealing with an "impoverished" ecosystem approach (in contrast to "evolutionary ecology").

Scientists in the ecology community were debating some of these issues, but in many cases, anthropologists did not fully understand the bases for the debates, plus the human dimension added profound levels of complexity to these issues. In any case, the emotion behind these critical writings and the use of such intemperate terms such as "reification," "vulgar," "obsession," "false," and "impoverished" reflected the intense feelings about ecological and cultural materialistic approaches by a majority of anthropologists.

Thomas's Work

The work that Brooke Thomas conducted on energy flow research in a highland native community in Peru was begun in the late 1960s after Rappaport's and Vayda's ecological studies of New Guinea populations. The work was stimulated by Rappaport's and others' research and by the ecologist H.T. Odum's graphic shorthand language to represent the flows and controls of energy through ecosystems. At the time of the study, Quechua natives of the altiplano employed a mixed subsistence of cultivation of potatoes (and other tubers) and quinoa (chenopods) and herding of llamas, alpacas, and sheep.

By comparing food energy production (outputs) with labor expenditures (inputs), Thomas demonstrated that cultivation provided a 10: 1 return, while livestock herding provided only a 2: 1 energy return (Thomas, 1976). Animal products (meat, hides, wool) were highly prized at lower elevations; hence, trade of animal products for other foods (e.g., maize, sugar) increased the ratio to more than 7: 1. Thomas's model, although representing averages and a simplified view of the energetics of production and expenditure in this community, nevertheless quantitatively demonstrated the utility of some of the principles of Quechua native subsistence through energy flow.

Thomas's work was the focus of an intense critique in a book called Energy and Effort that was edited by the distinguished human biologist Geoffrey A. Harrison (1982). The critique was penned by Philip Burnham, who began his comments by criticizing H.T. Odum's (1971) work on Environment, Power, and Society, identifying it as "reductionist" and "breathtakingly naive."

This book was somewhat naive, particularly in its chapters on human politics and religion, but many of the analytical approaches were very useful. Burnham (1982) continued his comments by outlining methodological problems that he saw as limiting understanding of human behavior by energy flow studies. One point has merit, where he stated: "...there is the problem of the multidisciplinary competencies required of a single researcher engaged in human ecological field study..." (Burnham, 1982: 233). Other arguments that he made were:

(1) that the costs are too high for the "pay-offs" of energetics (anthropologists have grown accustomed to very modest research budgets);

(2) adequate nutritional assessment is impossible from field studies (Michael Latham, an eminent nutritional scientist from Cornell University once told me that it was really the

anthropologist who could address several key nutritional issues from extended field work);

(3) too many simplifying assumptions were made (this is a key to modeling, but only at the outset);

(4) it is impossible to account for all of the social issues (but, this is never even possible in sociocultural analyses);

(5) it is inappropriate to apply the functional/adaptational paradigm borrowed from biology (this reflects the hostile views toward the biological sciences that many social scientists feel). Adaptation as a concept was criticized heavily where he expressed his view on "...the inadequacy of the concept of adaptation as applied to social behavior!" In brief, Burnham typified the views of many sociocultural anthropologists (despite his interests in human ecology) where a materialistic, adaptational, quantitative approach that draws on basic biological principles somehow sidesteps the fundamental bases of human culture and society.

CHANGES FROM THE 1980S TO 1990S

Ecological anthropology was popular within the anthropological community during the 1960s and early 1970s. This was also the period of the International Biological Programme and dramatic advances in ecosystems science, but these were largely separate events. In anthropology the ecological approach to understanding human social behavior soon began to fall out of favor, such that the late 1970s and 1980s were characterized by approaches that were antithetical to science. During the 20th century and into the 21st century there have been tensions within anthropology linked to different traditions and approaches to inquiry.

For sociocultural anthropology, the balance between scientific and humanistic approaches moved in favor of the humanistic in the late 1970s. "Critical anthropology," "deconstuctionist," Marxist, and "postmodern" literary approaches began to find their way into anthropology theory. Applied studies, within the realm of economic development in the Third World, began to increase. At the same time, in the United States many graduate programs in anthropology moved away from what limited scientific and quantitative training that they had received in the past. Biological anthropology during the second half of the 20th century had been the most scientific of the subfields of anthropology. But because the traditions of research were growing apart, biological anthropologists and scientificallyoriented sociocultural anthropologists were becoming marginalized within the anthropological community. All of these events affected the application of ecological principles to the solution of anthropological problems during that period. By the early 1990s, the position of human ecology within anthropology appeared to be moving in a positive direction. In a revised version of Emilio Moran's (1984) edited work on The Ecosystem Concept in Anthropology published by Moran in 1990, the introduction and papers that followed were forward looking and encouraging about ecological approaches to anthropology. Rappaport (1990), who contributed to this edition discussed the frequent shifts in theoretical perspectives that plague the field of anthropology:

> *"I both predict and encourage another swing of the pendulum. I predict revitalization of the ecosystem concept because it seems in accord with a general public's commonsence experience of a world beset by multiplying and interrelated environmental disorders, most of which it can attribute to humanity itself. I encourage this revitalization, with appropriate modifications, because the ecosystem concept itself is a vital element in the construction, maintenance and reconstruction of the webs of life upon which, by whatever name we call them, we are absolutely dependent." (Rappaport, 1990: 69)*

The anthropological pendulum did appear to be swinging back, but with a series of different approaches than those in the 1960s when Rappaport's first research was conducted. But before we deal with some of these newer ecological approaches in anthropology, let me discuss some of the theoretical issues that make it difficult for mainstream sociocultural anthropologists to accept ecological approaches within their field.

DIVISIVE ISSUES IN ANTHROPOLOGY (AND OTHER SOCIAL SCIENCES)

In 1968, Vayda and Rappaport stated that: "...a unified science of ecology has definite contributions to make towards the realization of anthropological goals and does not entail any appreciable sacrifice of traditional anthropological interests" (Vayda and Rappaport, 1968: 497).

There are a number of reasons why this objective has been achieved only in small measure that relate to some fundamental traditions in anthropology.

Following the Second World War, as with many other sciences, there was an increasing specialization by subfield accompanied by a tension between sociocultural anthropology and biological anthropology. Part of the basis for this was that in the 19th century and early 20th century, physical anthropology was preoccupied with race studies and there were clear racist elements in many of these studies.

This tension between "social" and "biological" intensified during the second half of the 20th century, when social scientists were concerned about "biological" and "genetic determinism" and other paradigms that placed heavy emphasis on human biological processes taking precedence over behavioral and social processes. Asuspicion by social scientists of all biologically based paradigms arose on the one hand and was paralleled on the other by a need to defend fundamental social processes and theory on the other. This "biophobia" is by no means universal in the social side of anthropology, but it does play an important role in the acceptance of certain ideas. Another tension that divides anthropology is the difference in approach between the scientists and the social humanist/historian (materialist interpretation vs. a symbolic/ cultural interpretation). Beliefs that human social behavior is so complex that is can never be fully understood by conventional scientific approaches are quite common among anthropologists, and, in fact, limit attempts to systematically study human behavior.

Some time ago, postmodernism entered anthropology via literary theory with challenges against the fundamental value of systematically-gathered information, and even objective reality. Within the social humanistic side of anthropology, there is a strong interest in "praxis" or practice or applied anthropology in reducing the effects of poverty in the Third World as well as in Western nations. Associated with these applications of anthropological knowledge is a kind of "anthropocentrism" (Rappaport, 1984: 387), which places the environmentalists and scientists who are concerned with the conservation of nature at odds with the social scientists, who see the world filled with poverty that has arisen in part because of differential knowledge and what is known as unequal power relationships. In this context, the social scientists are inclined to follow the idea of "putting people first," also the title of a successful collection of papers of development in the Third World (Cernea, 1991). This struggle between human needs and the need to maintain viable ecosystems is an exquisite conflict with an uncertain outcome (Newmark and Hough, 2000).

It is an area of investigation where collaboration between natural and social scientists is urgent! Another issue is an extraordinarily complex one: that of Garrett Hardin's "Tragedy of the Commons" (Hardin 1968). Agreat deal has been written about this issue, both within and outside of anthropology (Bollig and Schulte, 1999; Casimir and Rao, 1998; McCabe, 1990; McCay and Acheson, 1987). The fundamental objection that the anthropologists have against Hardin's basic premise is that it violates the concept of human agency; that is, the ability of humans to manage their own environments, and through cooperation, to avoid the tragedy of environmental exploitation that Hardin described. This belief in human agency is also linked to the unwillingness of many anthropologists to even entertain the idea that human behavior can be influenced by the circumstances of their environment.

An issue of the Social Science Research Council Items & Issues newsletter (Wissoker, 2000) was devoted to an article and commentaries on "advancing interdisciplinary research." In the lead-in to the collection, the editor noted: "...indeed, the idea of interdisciplinarity was practically born here..." What is significant about the contributions is that all of the commentary is by social scientists, and there is absolutely no mention of health, disease, the environment, ecology, or any of the natural sciences in their schemes of interdisciplinarity! These social science approaches to human ecology, are what in anthropology was called "cultural ecology." This approach, with an ecological emphasis on sociocultural process within the context of current anthropological theory was reviewed 25 years ago by Orlove (1980) and more recently by P. Little (1999).

Despite the bleak picture I have painted of social science and of anthropology in the context of science and ecology, some good work has already been done, and I am hopeful that new programs of collaborative and multidisciplinary research can be initiated between ecologists and anthropologists. However, it should be emphasized again that it is probably impossible for a single anthropologist or a single ecologist to conduct a study of human ecology and reach meaningful conclusions. The tasks are too vast for single scientists working alone and the solution is, of course, to establish multidisciplinary projects. Some examples of earlier and ongoing research and prospective ecological studies can be discussed.

INTEGRATED STUDIES OF SINGLE POPULATIONS

Beginning in the early 1960s, at the time that the International Biological Programme (IBP) was being organized, a number of single-population integrated projects were begun. Most of the projects were identified as a part of the Human Adaptability component of the IBP and were initiated with the concern that these populations were endangered, and their extinction would mean the loss to science of populations that most closely resemble human populations during the greater part of our evolutionary past. Later projects in the same pattern of investigation were started in the 1970s, some under the Unesco Man and the Biosphere (MaB) program. Human biologists or biological anthropologists organized most of these projects, but some were integrated with social scientists playing key roles.

There were several themes that these projects represented, including: adaptation to the environment, in its broadest sense; microevolution; cultural and biobehavioral evolution; health, epidemiology, and culture change; and ecology and systems science (Little et al., 1997). Some of these major integrated and multidisciplinary projects are listed here. Adaptation to (1) arid environmental conditions and limited resources in Kalahari hunter-gatherers; (2) high-altitude hypoxia and cold in Andean Quechua; and (3) Arctic cold in circum-polar Siberian, Inuit, and Algonkian populations, were studies conducted within a framework of populations living under the stress of extreme environmental conditions. Microevolutionary studies were conducted of the genetics of the Amazonian Yanomama, Makiretare, Cayapo, and Xavante, the Andean Aymara, the Central American Garifuna, Solomon Islanders, and several populations of central African Pygmies.

Language, genes, demography, culture, and phenotype, were used to explore ongoing evolutionary processes in these populations, and to reconstruct processes in the past. Attempts to reconstruct cultural and biobehavioral evolution of the paleolithic were made in the Kalahari and Pygmy studies. Health, epidemiology, and culture change were central issues in the Circumpolar, Tokelau Island Migrant, Samoan Migrant, and several other projects. Here the effects of modernization on native populations were a primary objective. Finally, although an interest in the influence of the environment on a population's behavior and biology was a common theme in all of these integrated projects, only a few had real interests in ecology and systems approaches (Little et al., 1991).

As already noted, Thomas (1976) carried out energy flow modeling on Andean Quechua farmer-herders. Later modeling focused on attempts to explain why some nutrient sources are important and why others are not. Gage (1986) applied optimal foraging theory to the slash and burn agriculture of the Samoans and identified that the net rate of energy production (NREP) was indeed a central guiding principle when Samoans considered production of their three primary crops: breadfruit, banana, and taro. Hett and O'Neill (1974) developed a carbon flow model for Aleuts that demonstrated a heavy dependence on marine organisms and the need to incorporate terrestrial and marine ecosystems in the analysis of Aleut food webs.

Finally, systems ecology figured prominently in the Kenya Turkana research because that was a central approach taken, but also because ecologists and anthropologists worked closely together (Coughenour et al., 1985; Ellis and Swift, 1988; Little et al., 1990; Little and Leslie, 1999). The synthesis of this work concluded that Turkana pastoralists were capable of surviving and flourishing in a dry and highly variable ecosystem by complex livestock management, mobility, opportunistic exploitation of resources, and adaptive social patterns of sharing-while at the same time, avoiding degradation of the ecosystem (Leslie et al., 1999)

CURRENT APPROACHES IN ECOLOGICAL ANTHROPOLOGY

There are a number of approaches to a human ecology that have been applied since the early 1980s. These represent the increasing specialization in anthropology, not only by the subfields that were described earlier on, but also by different theoretical approaches. Some approaches, parallel those taken in the field of ecology, but with time lags of several years.

Political Ecology

Political ecology is derived from political economy, in which there is concern with social inequalities and power relationships. It developed as a reaction to what some considered as an emphasis on ecological explanation for human social behavior to the neglect of political factors. The label has also been used in a Marxist context with the argument that "...an expanding capitalist economy is destructive to the environment." (Vayda and Walters, 1999). In this latter context, the ideas are principally Marxist in origin, not anthropological. Some anthropologists have criticized the contemporary application of political ecology as moving toward too much emphasis on "political" and too little emphasis on "ecological" relationships (Vayda and Walters, 1999).

Evolutionary Ecology

Evolutionary ecology arose from Mac Arthur's (MacArthur, 1960; MacArthur and Pianka, 1966) work in the 1960s that combined ideas from Darwinian evolution, ethology, population biology, and mathematical modeling. Much of the work deals with mathematical models of behavior within an adaptation framework. Anthropologists have been interested in this area of combined economic and ecological modeling of human behavior since the 1970s (Dyson-Hudson and Smith, 1978; Smith, 1979; Thomas et al., 1979). Sometimes the research is identified as behavioral ecology (Borgerhoff Mulder and Sellen,1994). Four areas of research that are germane to anthropology were identified:

(1) foraging strategies;

(2) mating systems and life-history strategies;

(3) spatial organization and group formation; and
(4) niche theory, population dynamics, and community structure (Smith, 1983).

Much of the research to date has focused on optimal foraging among hunter-gatherers such as the Peruvian Amazonian Piro (Alvard ,1995), Ituri Pygmies (Bailey, 1991), Paraguayan Ache (Hill, 1988), Canadian Inujjuamiut (Smith, 1991), and Canadian Cree (Winterhalder, 1983).

Other applications, such as optimal foraging of nomadic pastoralists have only been applied in a handful of cases (De Boer and Prins, 1989; Edwards et al., 1994). Borgerhoff Mulder and Sellen (1994: 225) identify the future of pastoralist studies as lying in "...a successful combination of quantitatively based studies and powerful modeling techniques," especially in the application of optimality models. Anthropologists are particularly well-suited to this kind of detailed observational research, because of lengthy time requirements for field observation within the tradition of extended field work in anthropology.

Historical Ecology

Historical ecologyis a relatively new approach in ecological anthropology that has been embraced by some archaeologists and ethnohistorians. Early works of interest to anthropologists were, among others, by Wiliam McNeill (1976) and Alfred Crosby (1972, 1986), historians who documented events linking human health and the environment in historical perspective. Based on a conference held at the New School of American Research in Santa Fe, New Mexico (Crumley, 1994), the field appears to be defined as a frame-work for studies of past ecosystems and their changes through time, with attempts to sort out the effects of anthropogenic and natural (non-anthropogenic) processes.

Most practitioners of this approach are from ethnohistory and archaeo-logy, and they build their theory on ideas from landscape ecology, geography, archaeology, history, and ethnohistory. This is an important application of ecology to anthropology, since losses in biodiversity during the present century can be placed in the context of earlier times through studies of prehistory. This will be discussed below in the context of biodiversity.

Landscape Ecology

Landscape ecology, with its background in geography and geomorphology, has a particular appeal to sociocultural anthropologists because of their current interests in land use in the Third World (Coppolillo, 2000). Archaeologists, as noted, are also drawn to this framework (in the context of historical ecology) for research because of the anthropogenic transformations of the landscape that are a part of human prehistory and history (Balée, 1998).

Ecosystems Ecology

Ecosystems ecology became the dominant research paradigm of the International Biological Programme (IBP), but still, the incorporation of human populations was limited (Worthington, 1975; Collins and Weiner, 1977). One of the problems with the Human Adaptability Component of the IBP was a conflict that arose between sociocultural and biological anthropologists during the early planning of the IBP. Margaret Mead argued for a largely social approach to the Human Adaptability research, but was voted down at an ICSU General Assembly. She then pulled out her support for the program and the anthropological community followed suit (Weiner 1977). This then led to a dominance of the Human Adaptability research by human biologists and biological anthropologists.

Later, two conferences were held in an attempt to coordinate some of the IBP biome research with human adaptability research projects, but with limited success (Little and Friedman, 1973; Jamison et al., 1976). There is still resistance among some biological scientists who have interests in ecosystems analysis to incorporate human biologists or social scientists in their research for some of the obvious reasons discussed above. However, more projects in the 1980s and 1990s, including those under the aegis of Unesco's Man and the Biosphere Programme, have successfully conducted collaborative ecosystems work on human populations (Boyden, 1992; Hladik et al., 1993, Little and Leslie, 1999; Lusigi, 1981). Collaborative efforts between ecologists, other natural scientists, and social scientists have been initiated in the urban LTER (Long Term Ecological Research) projects. The Central Arizona-Phoenix

LTER was codirected by an archaeologist, and the program is dependent on social scientists to develop realistic models within an ecosystems framework (Collins et al., 2000; Redman, 2005).

Ecology of Health and Adaptability

Studies of the ecology of health and adaptabilityof non-Western populations provide a breadth of environmental and health conditions not usually experienced by Western peoples. It is therefore important to study traditional as well as industrial peoples to gain insights into the full spectrum of environmental influences on health. Ecological and biogeographical approaches and models are often useful in understanding health threats and risks. One model that has been useful employs the movement of people from one environment to another to test for effects of the new environment on health.

An IBP project investigated the effects of movement of nearly 1000 Pacific Tokelauan Islanders to New Zealand after a disastrous hurricane struck their island in 1966. Baseline health data of Tokelauans collected in 1963 were compared with New Zealand Tokelauan health, and these migrants were found to have higher prevalences of obesity, type II diabetes, asthma, and hypertension than the native islanders (Prior et al., 1977; Fleming and Prior, 1981). Studies of migrant American Samoans to Hawaii demonstrated essentially the same effects of migration and modernization (Baker, 1984; Baker et al., 1986). The principal variables associated with declines in health have been identified as diet, activity, and levels of stress associated with a Western life style.

Ecology of Reproduction

The Ecology of reproduction is an area of interest that is central to human population ecology and bears on the ecological processes that influence human reproduction. Biological scientists have known for a long time that the environment, particularly the availability of resources, profoundly influences reproduction. The earliest interest in this area was by anthropological demographers, with training in biobehavioral sciences, who were willing to entertain the possibility that humans were subject to the same biological rules as other organisms (Ellison 1990). I recall the advice of my close colleague in social anthropology, Neville Dyson-Hudson, who had a deep interest in livestock. During the planning of the Turkana research, we agreed that if we could identify environmental effects on the livestock, then we should look for similar effects in the human population. This was a radical view and would have been scorned by our sociocultural colleagues, but turned out to be a very productive way to generate hypotheses.

It was particularly useful in Turkana studies of child growth, lactation and breastfeeding, maternal health, reproduction, and fertility (Gray, 1996; Little et al ,1992, 1993; Little and Leslie, 1999; Leslie and Fry, 1989; Leslie et al., 1996).

PRODUCTIVE DIRECTIONS FOR ECOLOGICAL RESEARCH IN ANTHROPOLOGY

We believe that the answer to the question posed in the title of an earlier talk on whether there is a future for ecological studies in anthropology is "yes," but probably still at the margins of anthropology, not in the mainstream. Carole Crumley (1998: ix-x), an archaeologist with interests in historical and landscape ecology, observed that the anthropological subdisciplines most closely allied with the sciences-archaeology, biological anthropology, and human ecology within sociocultural anthropology-have been marginalized in anthropology for much of the latter half of the 20th century. These are also the subdisciplines most closely allied with environmental sciences, and, hence, are more receptive to ecological approaches.

There are several areas of exploration within anthropology that would profit from ecological applications and multidisciplinary collaboration. These include studies of: landscape ecology within sociocultural anthropology, historical ecology within archaeology, urban ecology and managed ecosystems, within anthropology, broadly, and biodiversity and global studies, again, within anthropology, broadly.

Landscape Ecology within Sociocultural Anthropology

There is considerable interest in land use in the Third World by sociocultural anthropologists and other social scientists, particularly in the context of human population growth and increasing pressure on soil and land resources. There are disagreements over whether pastoralists and cultivators contribute to desertification, and how cultivation of the land can be best managed in semiarid or wet tropical lands. I recall an important observation that George Innis made in modeling slash-and-burn agriculture in the tropical rain forest more than 25 years ago. Innis (1973) reported that repeated use of tropical swidden plots, even with up to 40 years of fallow between each period of cultivation, would lead to depletion of several essential soil nutrients (potassium and soil organic matter), and that some of these nutrients required more than a hundred years for full recovery. What struck me was that conventional anthropological wisdom dictated that the indigenous pattern of eight years fallow was sufficient to maintain sustainability indefinately. The value of this pattern of use was embedded in the anthropological literature, yet it was clearly false. In this case, so much could have been gained by collaboration.

Historical and Landscape Ecology within Archaeology

Historical ecology, as a new paradigm, has been embraced by archaeologists with materialist and environmental interests. Since the landscape and landscape transformation are central concerns, the ecological conditions of human history must be understood within the context of the cultural conditions for a comprehensive interpretation. Some consider it "..the most important current intellectual advance in the study of human and environmental relationships" (Balée, 1998: 2).

I would moderate this statement a bit, but I agree that this approach does hold great promise for integrating archaeology, ecology, and history in interesting ways. Anthropogenic landscape alteration in prehistoric times is well documented, especially in the New World before Columbus. Terrace, canal, and road construction in the Andes produced dramatic transformation of the land-scape, and the lowland Amazon forest was modified by slash-and-burn cultivation and by raised fields. Raised-field cultivation was a widespread agricultural technique and used throughout the lowlands of Central and South America. A recent application of this kind of landscape archaeology is in the exploration of raised cultivation fields in the seasonally flooded areas of the Bolivian Llanos de Mojos (Erickson, 1995). Techniques used in the Llanos de Mojos work, in addition to archaeological methods, included agro-climatic modeling, ethnobotany, remote sensing, and experimental construction.

Few people use these raised fields today in South America despite their effectiveness and the apparent sustainability of this indigenous pattern of cultivation.

Urban Ecology and Managed Ecosystems

One of the dominant trends in human populations over the past century has been the movement of people from rural to urban settings (Bogin, 1988). Rural-to-urban migration takes place largely because individuals perceive that cities are centers of economic opportunity and excitement. The process, whether within or between national boundaries, has contributed to remarkable urban growth and widespread conditions of congested living that are unprecedented in human history. The beginning of the 21st century was marked as the time in which more than 50 percent of the earth's population were living in cities. Migration from the countryside to the city dates back to the rise of cities in antiquity (McNeill, 1978), and has been one of the most common types of migration since that time. Indeed, until the last century, urban mortality rates were so high that most cities could not even maintain their sizes, much less grow, without substantial numbers of immigrants (McNeill, 1979).

The linkages between rural-to-urban migration, demography, epidemiology, and urban ecology are crucial ones if we are to understand this highly-modified urban ecosystem. The urban LTERs in Phoenix and Baltimore, as well as other urban studies around the world, can serve as test cases. If the topical mix is demography, epidemiology, and urban ecology, the disciplinary mix should certainly include the social, biomedical, and ecological sciences.

Biodiversity and Global Studies within Anthropology

Within the past two decades or more, ecologists have become increasingly aware of losses in numbers and kinds of organisms around the globe and in alterations in the biosphere (Raven, 1997; Solbrig et al., 1992; Wilson, 1988). At the same time, climatologists and other scientists tracing global trends have identified disturbing patterns associated with anthropogenic effects on the planet (e.g., increased atmospheric CO^2 and pollution, progressive soil loss). Since anthropologists, particularly biological anthropologists, are more closely attuned to environmental effects on humans, they have been slow to become involved in exploration of biodiversity losses. Another limiting factor to the participation of anthropologists in assessing changes in the biosphere is the problem of spatial scale; that is, such large-scale, global problems are usually outside the scope of anthropological investigation.

One international program in which anthropologists and other social scientists might participate as members of multidisciplinary teams is DIVERSITAS, an international program of biodiversity research, consisting of 11 research components (Diversitas, 1996). The "Human Dimensions of Biodiversity" component of DIVERSITAS incorporates human-oriented disciplines within the general themes of the other components and is designed to contribute to an integrated approach to the study of losses in biodiversity. Most recently, the U.S. National Committee for the International Union of Biological Sciences (IUBS) prepared a document that defines how some social scientists and human biologists might contribute to DIVERSITASefforts (Little et al., 2001). The U.S. National Committee for the International Union of Anthropological and Ethnological Sciences (IUAES) will also contribute to the definition of a U.S. Program in this area.

There are several essential areas in which anthropologists might contribute to these very important studies. First, there is the area of human impacts on biodiversity, which is certainly the central feature of biodiversity losses. The impacts of human populations in transforming the landscape, competing for habitats, contributing to pollution, and outright predation, are profound, indeed. Among many kinds of studies, it is here that archaeologists and ethnohistorians can document some of the long-term changes in biodiversity and how rates of loss have varied through time. Second, how has human biodiversity changed in the context of other species' losses? This is a vast area of exploration, but one that might draw human biologists into DIVERSITAS research. In a third area of exploration: how do human perceptions of biodiversity influence our abilities to respond to losses or to take action? How does culture play a role in these responses?

Finally, there are the interactive dynamics between human diversity and the environment, that is, the influences that losses in biodiversity will have on human health and well-being, including human social health (Hauser et al., 1994). These are all problems for the future that may or may not be soluble, particularly since solutions depend not only on technology but on human behavior, as well. Such behavior is dependent on a complex mix of culture, politics, economics, health, psychological perception, and many other categories of variables. If these problems are soluble, it will only be through integrated and multidisciplinary investigations in which social scientists join forces with natural scientists in this challenging endeavor to preserve our biosphere, including the species, Homo sapiens.

7

Tools and Methods for Historical Research

INTRODUCTION

The objective of this working paper is to provide some examples of the type and range of digital tools (and associated methods) that might profitably be used by researchers working in the discipline of History. In keeping with the remit of the Methods Network-one of the objectives of which is to promote the dissemination of knowledge about ICT related tools and methods across subject boundaries-some reference will also be made to techniques and tools normally associated with other disciplines which clearly have applications to historical research.

On a theoretical map of arts and humanities subject areas where disciplines could be positioned to try and demonstrate overlap and methodological common ground, there would be a strong case for placing the discipline of History squarely in the centre of the map, providing as it does a chronological backbone for all areas of research, as diverse as musicology, practice-based art, theology, literary studies and so forth. In terms of immediate adjacency, the borders of what might be considered 'historical research' would blur imperceptibly with a number of neighbouring subject areas: archaeology, classical studies, sociology and art history for instance; all of which have tools and techniques associated with them which are (or potentially could be) of use to historians. Appropriated techniques from computing and information science such as data mining and e-Science related methods are also increasingly coming to the attention of those seeking to explore new methods of engaging with research questions-as are techniques associated with geographical mapping and information systems-so it is clear that the scope for employing a variety of technical methods is very broad.

This view is echoed in a recent report written by Onno Boonstra and colleagues who outline numerous methodological and thematic areas of potential research, many of which they feel have yet to be tackled in a coordinated and purposeful way. Whilst very useful as a roadmap for possible engagement with a wide variety of technological tools and methods, the wider context of the report is gloomy in its overview of the recent state of 'historical information science' (the authors' preferred term for computer-related historical research).

At the time of writing in 2004, they expressed the view that the results of nearly two decades of History and Computing were 'slightly disappointing' and went on to qualify this statement as follows: They are not disappointing because 'computing' failed to do what it intended to do, which was to provide 'history' with computerised tools and methods historians could use to expand the possibilities and to improve the quality of their research, but because 'history' failed to acknowledge many of the tools 'computing' had come up with.

One of the major challenges that the authors identify is the provision of an effective internationally accepted forum for discussing and disseminating the results of this kind of research and it is clear that the various iterations of the Association of History and Computing (ACH) have a critical role to play in coordinating and promoting these kinds of activities, both at international and national levels.

TOOLS AND WEB RESOURCES

To address the question of what is currently available to historians, one must inevitably turn first to the Web for an overview of initiatives that are the result of recent (and not-so-recent) attempts to use digital tools to break new ground in areas of historical research. One early and very influential project that is widely cited as providing a benchmark for presenting complex historical research material on the web is the 'Valley of the Shadow', a hypermedia archive about northern and southern communities during the American Civil War containing more than 100,000 items taken from newspapers, letters, diaries, photographs, maps, church records, population census data, agricultural census data and military records. Hosted by the Virginia Center for Digital History, this project might usefully represent the medium of the web itself as one of the most ubiquitous and most widely accepted tools, complementing the use of email as the other major communication tool that presumably all historians now accept as an intrinsic part of their activities. A number of repository/gateway/portal/listings sites exist to aid historians in finding relevant resources.

With such a plethora of very effective portal-type tools for finding what amounts to an enormous array of resources for historical research, it is useful to have a recent AHRC funded report entitled 'Peer review and evaluation of digital resources for the arts and humanities'

The top ten resources listed were as follows:

Resource	Responses
Oxford Dictionary of National Biography	122
Royal Historical Society Bibliography	77
Early English Books Online	58
Eighteenth Century Collections Online	28
Times Digital Archive	24
Archaeology Data Service	19
The Proceedings of the Old Bailey	18
British and Irish Archaeological Bibliography	12
British History Online	11
Canmore	10
Hisiorical Abstracts	10
Thesaurus Linguae Graecae	10

Table : Taken from Peer Review and Evaluation of Digital Resources for the Arts and Humanities (pg.5)

Whilst these 'resources' are slightly to one side of the main thrust of this paper-the principal purpose of which is to consider 'tools'-it is nonetheless illuminating to understand which sites are perceived as particularly valuable. In addition to relevant content which will almost always be the principle reason that users will keep on returning to specific websites, a less important but nonetheless significant enticement will be:

- o The functionality that is available (e.g. search and sort options)
- o The flexibility of the data (e.g. variant spelling detection; text hyperlink to original manuscript image)
- o The robustness and durability of the data (e.g. the technological platform; the encoding standards employed)

Taking some of the above resources as examples, it may be useful to consider some of the tools associated with their development, functionality or implementation to understand what elements other historians might usefully aim to employ in their own ICT-related resource development projects.

The Oxford Dictionary of National Biography uses a very flexible search interface which offers a 'type and hope' simple search field for finding a specific person or word from the full text of the DNB, but also offers five other search screens incorporating many more fields (sometimes linking to further 'advanced' options e.g. name search) which allow the user to access all manner of gender related, professional, geographical, chronological, religious and financial information-in addition to image and bibliographical data. One of the comments in the survey pointed out the specific usefulness of this site in creating prosopographical statistics and it is clear that the online DNB fulfils the criteria often quoted for establishing the legitimacy of a technological resource, i.e. that it allows researchers to access information that would be impossible or very difficult to have discovered before that resource became available.

Obviously, one of the critical tasks in assembling a resource of this magnitude (55,800 lives; 63 million words;10,300 portraits) is the rendering of the original text into a machine-readable format, and whilst this is less of a problem for text that has incrementally been transferred into digital formats over the years (which may have been the case with the DNB) it becomes a huge technological challenge when major repositories of primary source material are only in printed formats and require either manual text transcription or some form of scanning procedure (image and/or optical character recognition) to transform them into digitally accessible archives.

The Eighteenth Century Parliamentary Papers Project, one of the Initiatives Funded by the Joint

Information Systems Committee (JISC) digitization programme, will use very fast bulk scanning equipment that is capable of scanning 600 pages per hour, facilitated by vacuum enabled page turning technology and laser guided edge detection sensors. The size, weight and cost of this equipment (see fig.1) clearly requires a long term institutional commitment and will exceed the strategic requirements of many organisations and projects, but it is symptomatic of a new approach to resource creation that such tools now exist that can process around one million scanned pages a year, using techniques that are sensitive to the fragile nature of the primary sources and mindful of the technical standards intrinsic to sustainable resource creation. At a recent Methods Network workshop, one participant reported that the Open Content Alliance is using this kind of hardware in a multiple-machine facility that perhaps begins to approach the sort of capabilities available to commercial entities such as Google and Microsoft, who are separately in the process of creating colossal digital repositories of books numbering millions of items.

Whilst different in scale to the commercial initiatives referred to above, Early English Books Online (EEBO) is an important archival record of virtually all books produced in Great Britain, Ireland and British North America before 1700, and also features prominently in the AHRC Peer Review Project. The relevance to historians of this archive extends to an enormous range of subject matter that encompasses royal, governmental and provincial official public documentation as well as a prodigious amount of social historical material relating to all classes of society. Whilst the image scanning of book pages in this project is of enormous significance, it is the related Text Creation Partnership (TCP) initiative that is of further interest in the present context. EEBO-TCP is managed by the Universities of Michigan and Oxford and financially supported by over seventy other institutions to produce manually keyed and SGML/XML encoded text editions of a significant portion of the EEBO corpus. The aim is to create full text editions of about 20% of the works represented by images in the archive, thereby enabling search functionality that goes way beyond keyword-type linguistic analysis processes (e.g. concordance, collocation and clustering), highly useful as those are for historians as well as those involved with diachronic linguistics. As a result of the textual encoding though, searches can also be formulated that, for instance, only search for proper names, or only search in the marginalia, or perhaps only search for non-English terms that appear in stage directions.

The decision by the Text Creation Partnership to rely on laborious methods of manual keying and encoding clearly has a critical effect on what it is possible to achieve given certain funding criteria, and is in contrast to the abovementioned Eighteenth Century Parliamentary Papers project which declares on its website that Abbyy OCR (optical character recognition) software will be used in conjunction with the Agora Content Management System to automate as much of the data management as possible. Given that there is no single recommended solution for major projects like these, it is a useful reminder that the best strategic methodologies and the most appropriate tools need to be defined according to the nature of the work being undertaken.

This might encompass cases where a hybrid solution involving automation combined with human expert feedback into the system is the preferred methodology, improving the chances of more accurate scanning as the project progresses and as the system learns from its mistakes. The Gamera project is an open source initiative to supply a programming framework for building recognition systems for 'difficult' historical documents. It is envisaged as a tool that programmers will use in conjunction with subject experts and supports a reiterative test and refine model, a graphical user interface and a modular plug-in architecture that is capable of performing five separate document recognition tasks:

- o pre-processing
- o document segmentation and analysis
- o symbol segmentation and classification
- o syntactical or structural analysis
- o output

Gamera stores output in an XML-based file format, which enhances its interoperability with other systems and is also in keeping with its open source ethos.

Database Structures

Of all of the computer-related tools for historical research, database (and statistics) packages have had the most consistent use by the widest constituency of researchers using digital methods in the discipline. In general terms, approaches in the past generally focused on information originating from social science research and the data that resulted from this work often took the form of lists. As such, the manipulation and analysis of this data (e.g. census records, state finance data, demographic data, mortality data etc.) leant itself primarily to being handled by database packages and there is a considerable amount of literature devoted to discussion of the different approaches to structuring data and the use of relational database management systems (RDBMS), such as dBase , Paradox, Clipper, FoxPro and Microsoft Access. One of the principle areas of discussion within this area was how to most usefully represent historical data and much of this debate seems to have focused on the relative merits of a 'source oriented' versus a 'model oriented' approach to structuring the information.

The 'model-oriented' structure required a great deal of preparatory data analysis and provided logical organizational entities into which defined data elements could be deposited (i.e. database fields). At the outset of a project then, not only was the researcher obliged to work out a sensible way of dividing the information into rational and discreet entities, but he or she also needed to think very carefully about what sort of information would need to be extracted from the system during its working lifecycle. This model was particularly effective in cases where the data was relatively simple, complete and regular in its structure.

The 'source-oriented' approach attempted to more comprehensively represent the source information in its original form using a combination of textual markup and fielded data structures, thereby capturing as much of the subtlety and nuance of the original data as possible and also enabling future researchers to formulate their own retrieval and analysis approaches to the information. As a leading exponent of the 'source-oriented' camp, Manfred Thaller developed the KLEIO system which represented the original source data at two levels: firstly as uninterrupted strings of arbitrary characters and secondly, as meaningful units such as numerical or calendar data etc. The more formally structured element of the system, arranged hierarchically and referred to as the knowledge or logical environment, accurately referenced the transcription (i.e. the full-text) layer, and all queries made of the transcription layer were interpreted via that logical environment-the net advantage being that the researcher still had a 'machine-readable' version of the original to refer to.

Whilst contextually useful for talking about data modelling, the debate about 'source' versus 'model' oriented approaches-which very broadly speaking might translate into 'encoding' versus 'database'-has diminished in importance due to the potential of XML query processes (e.g. Xpath and XQuery) and the relatively recent acceptance of the term 'XML database' to denote data in XML format that is interoperable with relational database packages such as DB2, Oracle and Microsoft SQL Server.

Where the functionality of a database is required, projects will undoubtedly continue to incorporate them into their technical strategy, but after a long period of standards being discussed intensively, it is clear that acknowledgement of some form of XML data interchange mechanism is almost a sine qua non of project funding, mainly because of concerns about preservation issues but also because of the growing awareness of the need to have an effective way of globally harmonizing disparate information systems.

Data Mining

Having considered issues to do with the presentation and structuring of data, the former principally in the form of web resources and the latter as database and textual information systems, the next stage of the information cycle involves the querying and analysis of that data. One deceptively simple experimental project, H-Bot, is an example of a 'question answering' (QA) system which accepts natural language questions and attempts (in one of its modes) to return exact answers using the

Google index to find statistically likely matches on the key keywords that the user has entered. The system is capable of performing reasonably well in the context of a certain type of factual question and its creators argue that further development work on the underlying rule sets would further improve its capabilities. Designing QA systems, according to Dan Cohen, 'exercises almost all of the computational muscles', involving as it does: search methodologies, document classification, question interpretation (natural language processing) and statistical and linguistic text analysis.

Cohen has also used a similar approach with another data mining system called Syllabus Finder which uses a keywords-in-context (KWIC) approach to return highly relevant results relating to syllabi information. Having determined a relevant keyword set using word frequency analysis from a known set of documents related to syllabi, the search term entered by the user is optimised by the addition of these terms and the bundled query then interrogates Google's API service (as well as a locally stored database) which returns search results in a SOAP envelope (an XML schema used for server-to-server communications). Additional statistical and expression matching analysis is then carried out on this dataset resulting in highly relevant and specific output that is capable of identifying college or university organisation names and course assigned book titles.

These and other tools created at CHNM are complex in their conception but are designed with simple interfaces and have the feel of small group developmental research systems that have been constructed to address particular user-defined needs.

A more widely collaborative data mining initiative that is currently focused on information retrieval in the area of literary studies is the NORA project, the objective of which is 'to produce software for discovering, visualizing, and exploring significant patterns across large collections of full-text humanities resources in existing digital libraries.'

The development of this software is being done by a consortium of organisations involving subject specialists and computing scientists and the functionality of the demo version of the system displays a similarly distributed methodology. The user interface software runs as a Java Webstart application delivered to the user via one server; the relational database management system (Tamarind) runs on another server; and the D2K data-mining framework currently runs on a third server. The D2K (Data-to-Knowledge) framework supports other projects: T2K (Text-to-Knowledge) and M2K (Music-to-Knowledge), all of which are being developed to play a significant new role in how the Humanities Computing community works together to build, use and share tools.

The principle is that 'modules' of well defined reusable code can be implemented as single or nested components to carry out a wide variety of functions-which in the case of the T2K framework would include a rich set of natural language pre-processing tools which could carry out: lemmatization, tokenization, part of speech tagging, data cleaning and named-entity extraction.

In the context of the NORA project, the concept is to build an application using the D2K framework which will retrieve 'only what's needed' from large digital repositories (of XML encoded text) and place that information into a database, thereby cutting down on the processing overhead of having to deal with large amounts of redundant or null data. Once collected, an additional function of NORA is then to provide visualizations of the analysis of that data that will provide researchers with clearer insights into a range of issues including social networks, content overview, document classification types etc. A project based at Sheffield University is also looking at the problem of mining information intelligently from distributed data sources and is specifically addressing how to deal with the sort of fuzzy and ambiguous data that often characterises historical information. The ARMADILLO project is based on twelve datasets that contain information about eighteenth century London and the objective is to employ probabilistic disambiguation methods (using pre-determined algorithms) to the heterogeneous data, which is then mapped to a fairly simple ontology consisting of the following entities:

- Name
- Role
- Place (i.e. place name which may be subject to a change of location)

- Location (associated with a map reference)
- Time (either point in time or time period)
- Resource

Users would then be able to turn these categories 'on' or 'off' depending on the nature of their search and the likelihood of returning meaningful results, allowing for much finer control of the data retrieval process. The stated aim of this project is to employ as much 'knowledge' to the task of initially processing information as possible, in order that the likelihood of error introduced by synonyms, homonyms, variant historical spellings, inconsistencies introduced by human error-and so forth-are minimised when records are retrieved for analysis.

Quantitative Methods

The use of quantitative methods is an intrinsic part of historical research but has not always met with blanket favour over the years in all areas of the discipline. In a recent essay, William Thomas cites the controversy over the publication in 1974 of Robert Fogel and Stanley Engerman's Time on the Cross: The Economics of American Negro Slavery which (in amongst a vast amount of other detail) posited the claims that only two percent of the value of income produced by slaves was expropriated by their masters and that the typical enslaved person received less than one whipping per year. These arguments were based on economic models and mathematical methods of analysis (cliometrics) and unsurprisingly met with fierce criticism from many quarters.

Despite this sort of opposition, however, the use of statistical techniques is deeply embedded into economic and social history research, as well as being a central component of data mining. In journals such as: Social Science History; Social History; and Economic History Review, accounts of the methodological processes employed in the various analyses featured are subordinate to the factual historical substance of the article and may therefore be giving a misleading impression of the acceptance and use of these techniques in various sections of the historical research community.

The most widely used statistical tools are borrowed from the Social Sciences where there seems to be a consensus that the SPSS system, first developed in 1968, continues to offer the correct balance of usability and functionality for most forms of analysis. Minitab software is also widely referenced by course descriptions for academic departments offering applied statistics modules, as is SAS, despite the fact that there are also numerous freeware and open source options available, many of which claim to focus on specific areas of functionality.

The ways in which statistics packages can be used is a complex and highly specialised field of study in its own right, but Boonstra et al propose the following methods as holding 'great promise for future historical research.'

- Logistic Regression-a model for predicting a dual category variable (e.g. married/unmarried, live/die, north/south
- Multilevel Regression-an analysis that includes hierarchical data (e.g. data about patients, the doctors treating them, the hospitals they work in, and the region the hospital is located)
- Event History Analysis-a study of the independent variables which may contribute to the likelihood of an event occurring
- Ecological Inference-reconstructing individual behaviour from aggregate data where there is a paucity of information relating to the individual
- Time Series Analysis-a range of techniques for studying change over periods of time Boonstra et al contend that the adoption of these techniques across all areas of historical study would enable interesting and productive research.

A related area where quantitative tools are widely used is in the construction of simulations, another technique that has been in existence for a very long time but has not been taken up consistently across the community.

The purpose of this technique is to analyse behaviour and events in the context of historically given variables in order to gain a better understanding of cause and effect within a defined system. Very old models include the semi-computerised attempt to simulate the outbreak of World War One by Hermann and Hermann (1967), and also SOCSIM (1970 onwards), a demographic micro-simulation program that enabled probabilistic analysis of marriage and fertility patterns within a closed society. Despite the lengthy history of their use in historical research, the application of current simulation techniques largely appear to be associated with risk assessment for businesses, environmental planning and scenario reconstructions for military situations.

Markov Chain sequences, which are widely used in simulation models and are relevant to determining the probability of an event occurring based on the occurrence of a previous event, have also been very influential in the field of Linguistics where much research has focused on probability models for automatically tagging large corpora.

It should be apparent that the potential for corpus linguistics techniques to impact on historical research is significant and that problems such as the disambiguation of pre-modern-era spelling variants (which is the focus of the VARD project at the University of Lancaster and a component part of the Historical Thesaurus of English) should not require duplicate research effort outside of the Linguistics community.

Visualization

Charles Joseph Minard's visualization of the Napoleonic Army's advance and subsequent retreat from Russia in the 1812-13 campaign is widely cited as one of the most effective graphical depictions of data drawn from statistical sources (see Fig. 3). The thickness of the line indicates the strength of the army during its peregrinations back and forth across Europe whilst the lower graph maps those fluctuations against the recorded temperature on the return leg of the march.

The clarity of information that this graphical display imparts is an impressive model of what visualization can achieve and is a useful benchmark for the type of digital output that it might be possible to create for various forms of historical data.

The Visual Spatial Technology Centre (VISTA) at the University of Birmingham is one of leading centres in the UK for visualization research and one of the many activities that it supports is the Medieval Logistics project.

During a recent Methods Network meeting, Helen Gaffney laid out some preliminary ideas for a visualization of the catastrophic defeat of the army of Emperor Romanus IV at Manzikert in 1071.

Using data from: settlement land use; known historical logistics issues and ecological, environment and terrain data, she proposed a vizualisation that encompassed a variety of aspects of decision theory-including network, game and optimal foraging theories-in order to elucidate the defeat of the Emperor's army at the hands of the less numerous Turkish forces.

Other initiatives in development at Birmingham include a project to investigate and visualize how drapery and adornment in the classical period would have affected movement and interaction between figures; and another demonstration involved the 3D visualization of the contents of a Canopic jar, which is thought to contain a human liver.

This kind of work marks out territory shared by historical studies and the cultural heritage sector and in terms of the cross-disciplinary sharing of tools, it is clear that methods employed by curatorial and conservation professionals, not least in how to describe objects effectively (using tools such as the CIDOC-CRM ontology), could be of enormous value to historians. The Wroxeter Hinterland Project, also based at the University of Birmingham, is a useful example of a project that employed an arsenal of tools to interpret the remains of a Romano-British urban archaeological environment.

An international team of archaeologists is carrying out the total exploration of the city by non-destructive remote sensing methods, including magnetometry, resistivity, electrical imaging, seismic scanning, ground probing radar, airborne hyper-and multi-spectral scanning, and satellite imaging.

Archaeology as a discipline is generally credited as having had more exposure over a longer period to technological means of research than most other arts and humanities subject areas and as such, represents an enormously useful pool of expertise for colleagues in neighbouring areas of historical research who are looking for new ways of analysing data in their area of research.

Another centre of expertise is the King's College Visualisation Lab, who are involved with a range of 3D visualization projects that attempt to map internal and external spaces using historical data sources to define verisimilitude as far as possible with the original environments.

Using 3D Studio Max and other tools, stunning results are possible that give fresh insights into how historical spaces were perceived and used and a recent initiative called the 'London Charter' is an acknowledgement that the visualization community requires standardised methods for documenting working practices.

At a more accessible level for the individual scholar, devices such as timelines, lexis pencils, graphs and cluster dendrograms are all examples of data visualization tools.

Geographical Information Systems

To find a remarkable implementation of a system that uses GIS as one of the component parts of its technology, one need look no further than Google Earth to illustrate an essential facet of what it is possible to do using such techniques. Satellite imagery of actual terrain can be overlaid with a wide variety of information choices including transportation, roads, shopping and services information and even user generated information about very specific buildings or neighbourhoods.

Obviously, the type of technology that Google has the capability to implement has global significance and requires an extraordinary amount of computational and storage capacity, but nonetheless, the layered approach to information is the same principle that arts and humanities academics (principally perhaps archaeologists) have used for many years when applying GIS principles to research.

The basic unit of representation in a GIS system is the layer, which has information associated with it to describe what that layer represents. Spatial data refers to the actual point on the earth's surface that the point or boundaries of the layer refer to and this is described using either latitude and longitude readings or other co-ordinate systems such as the British National Grid or the Universal Transverse Mercator (UTM).

The boundaries of the layer for this purpose can be defined as either a point, a line, a polygon, or a pixel, and each of these is represented using co-ordinate pairs. Attribute data constitutes the information that is associated with the spatial data and might consist of any information, but is generally text or statistics or graphical representations of data that is relevant to that location.

The industry standard package is the ArcGIS suite of tools which incorporates ArcView and ArcInfo, widely used by academic and commercial project teams, but a range of other products are available, including a freeware system called GRASS (Geographic Resource Analysis Support System).

It is clear then that GIS has enormous value for the annotation of maps, but it is has perhaps been less clear to historians over the years how a temporal component can be introduced to the system that will allow the description of changes over time.

At a recent Methods Network Expert Seminar, Ian Gregory gave a presentation that referred to problems of 'time' and 'space' and did so in the context of talking about infant mortality trends in England and Wales from the 1850's to the 1900's.

A temporal component to his research was provided by the Great Britain Historical GIS (GBHGIS, Gregory et al 2002), hosted at the University of Portsmouth, which provides census reports; data on births, marriages and deaths; and on unemployment and poor law statistics, for the entire period of Gregory's study.

The present incarnation of this system (in development since 2001) has provided scholars with a new resource in this particular area of research, and has also been the data source for a major lottery-funded project of wider immediate benefit to the public, the Vision of Britain website.

By overlaying statistical representations of infant mortality figures onto precise areas, it is possible to graphically display-in much more detail than was possible before-changes throughout the period and how they relate to a number of factors; notably the difference in death rates throughout urban and rural areas. A similar initiative, also launched in 2001 is the China Historical GIS, hosted at Harvard, which features historical maps and datasets that are freely available for academic use.

The use of GIS as a 'hub' technology around which an archive can be built is described in the AHDS Guide to Good Practice: A Place in History: a guide to using GIS in historical Research.

It quotes an example from the Perseus Digital Library, which features the Edwin C. Bolles Collection of the History of London. This includes maps of London from different periods presented as different layers, over which place names can be compared and references made to source texts.

Returning to the first project mentioned in this paper, the 'Valley of the Shadow', GIS is also a component of the functionality of this project and over 2000 individual dwelling places, taken from a very detailed map of Augusta County, Virginia (1870), are precisely pinpointed for the purposes of allowing users of the system to be able to identify the houses of particular individuals featured elsewhere.

CONCLUSION

By looking at the various parts of the information cycle throughout this paper, starting perhaps perversely with presentation; going through data structuring and modelling; onto query and analysis; and then back to presentation again, it has become clear that although the selection of tools, processes and resources referred to has been highly selective, they are symptomatic of a great richness of digital techniques that can be exploited by historical researchers.

As Greg Krane stated at a recent Methods Network workshop, Thirty years ago the dog was the print world and the tail was a few ancillary digital tools to augment that. I would say now the dog is the digital world and the print world is the tail.

In a discipline so dependent on the gathering and analysis of written sources, those doubting the efficacy and authority of digital techniques may find themselves in grave danger of being left at the periphery of historical research.

The references and examples used in this paper include work that has been presented or referred to at a number of Methods Network sponsored events, all of which have some material relating to them on the Methods Network website. In the context of workshops and seminars, a report is produced by the organiser of the event which summarises and draws conclusions from proceedings. Where the event has been designated an 'expert seminar', more detailed material such as draft papers, slides or audio recordings of presentations is made available. A list of Methods Network funded events that have broad relevance to historical research can be found at the end of the following 'reference' section.

8

Subsistence Strategies and Early Human Population History: An Evolutionary Ecological Perspective

Since the mid-1970s, an in uential view of human forager population history has been that foragers limit fertility in a way that maintains their populations in dynamic equilibrium with available resources (Dumond 1975; Hayden 1972, 1986). Within this framework it is generally assumed that human populations grow to some limit set by the carrying capacity of the environment and stabilize around that limit, with minor uctuations occurring due to density-independent factors affecting productivity. When productivity (i.e. the amount of energy acquired per unit area of land) increases due to intensi cation, this limit also increases, resulting in a new equilibrium. Thus, with the adoption of domesticates and attendant sedentism and increases in total production, birth intervals decrease, fertility increases and average rates of intrinsic growth start to rise.

The most commonly accepted estimates for population growth rates (Hassan 1981) during the Palaeolithic are based on assumptions regarding Palaeolithic forager population density, the amount of the earth's land surface actually occupied by humans in early prehistory and the population of the earth at the end of the Pleistocene. However, anyone with a hand calculator, the formula for exponential growth and only the vaguest notion of how many people were alive 12,000 years ago can quickly determine that average growth rates through much of Pleistocene human prehistory could never have been much higher than .01 per cent, an increase of only one person per 10,000 per year. Yet, during this period, virtually the entire habitable surface of the earth was colonized by humans in what amounts to little more than the blink of a palaeontologist's eye. During the Holocene, a period which coincides with global warming, the wide-scale adoption of domesticates and increased sedentism, these near-zero rates must have begun to rise, resulting ultimately in the population of a little over 6 billion in which we nd ourselves immersed today. Why were growth rates so low when essentially all humans alive were foragers? Why did growth rates increase during the Holocene?

The starting point for an alternative view of human population in prehistory in this article is Hill and Hurtado's succinct observation that 'No natural fertility population yet observed is characterized by zero growth, as would be required over much of our species' history' (1996: 471). Over the past decade or so, a number of empirical studies and theoretical treatments have appeared in the eld of evolutionary ecology and related elds of evolutionary anthropology that have begun to paint a different picture about what population dynamics were like in human prehistory and how they relate to broad changes in subsistence modes.

There has been a greater appreciation of the idea that individual energetic ef ciency in resource acquisition and production, rather than total productivity rates or the carrying capacity of the environment, play a critical part in determining reproduction rates (Belovsky 1988; Hawkes and O'Connell 1992; Winterhalder et al. 1988). Human foragers are no longer seen as the natural resource conservationists they once were (Alvard 1998). Closer analysis of extant human foragers has shown that their modal fertility differs surprisingly little from that of horticulturalists and intensive agriculturalists (Campbell and Wood 1988; Bentley et al. 1993a, 1993b). There has been greater appreciation of the fact that effective control of fertility has been quite rare in human history, and may be limited to special circumstances surrounding the recent demographic transition. Finally, the idea that human population history has been characterized not by a series of stepped dynamic equilibria but rather a saw-tooth pattern of periods of rapid growth interrupted by infrequent but serious crashes has become increasingly recognized as an alternative explanation for near-zero growth through much of human prehistory.

This in turn has interesting implications for understanding the selective environment that gave rise to human life-history characteristics, subsistence strategies and social structure (Blurton-Jones et al. 1999; Boone and Kessler 1999; Hill and Hurtado 1996: 471-2; Keckler 1997), as well as for changing rates of cumulative cultural evolution during the Upper Palaeolithic and the Holocene (Shennan 2000, 2001).

One of the keystones of the evolutionary ecological approach is the concept of energy budget, in which time and energy allocation is conceptually divided into somatic effort (growth, development and maintenance, including subsistence activities) and reproductive effort (which is further divided into mating effort and parental effort). Time and energy allocated to one component must be traded off against allocation to another. Using this energy budget approach in conjunction with some of the general implications of foraging theory, this article will explore the relationship between population dynamics and subsistence intensi cation. My discussion will revolve around two basic propositions regarding long-term human population history:

(1) the near-zero growth rates that have prevailed through much of prehistory are likely due to long-term averaging across a periods of relatively rapid local population growth interrupted by infrequent crashes caused by density-dependent and density-independent factors; and
(2) broad changes in population growth rates across subsistence modes in prehistory are probably best explained in terms of changes in mortality due to the dampening or buffering of crashes rather than signi cant increases in fertility.

THE DIET BREADTH MODEL AND ITS IMPLICATIONS

In any environment in which humans nd themselves, there is typically a wide array of animal and plant food items that could be successfully captured, collected, processed and eaten. And yet, rarely is it the case that human populations capture, collect and consume everything that available. In some contexts, human foragers tend to ignore small mammals, reptiles and birds, while, in others, such prey are pursued and consumed. Plants foods that are relatively time-consuming to collect and process, such as acorns or other seeds, are ignored by some foragers, and routinely collected, processed and consumed by others. What kinds of factors affect how humans choose which food items to pursue, process and consume?

The diet breadth model (see Winterhalder and Smith (2000) for a recent review of applications) was developed and employed to answer this kind of question. The logic of the model begins with the observation that different kinds of potential food items in the environment vary in the amount of time it takes to locate, capture or collect them, and to process or render them into a form available for consumption and digestion. Which ones should a forager expend time and energy capturing or collecting when they encounter them, and which ones should they ignore in favor of continuing the search for energetically valuable prey? To answer this essentially economic problem, the model makes several simplifying assumptions.

First, foragers encounter potential food items in the environment at random. Second, foraging costs are measured in terms of time, and total foraging time is partitioned into two mutually exclusive categories: search time, or time spent locating prey items in the environment, and handling time-time spent pursuing, capturing, processing and consuming the prey once it is encountered. Third, foragers rank their preference in potential prey in terms of pro tability, de ned as the net energy return rate obtained per unit of time expended in handling time upon encounter. The question then boils down to this: as a forager is searching through the environment and encounters a potential food item, should she pursue it and eat it, or should she ignore it and continue to search for something more pro table? The answer is: foragers should take lower-ranked prey only as long as the return rate per encounter (pro tability) is greater than the average return rate gained from searching for and handling higher-ranked prey.

One widely recognized implication of this is that high-ranked prey should be taken when encountered no matter how rare they become, and that prey ranked below the optimal diet breadth will not be taken no matter how abundant they are. This means that, given any population growth at all, depletion of foraged resources is nearly inevitable, and that depletion should begin with the highest-ranked resources. How low on a ranked list of potential prey a forager will be willing to go depends upon the encounter rates of higher-ranked prey. Thus, as higher-ranked resources are depleted, lower-ranked items are added to the diet. Lower-ranked prey are ranked lower because they entail higher handling costs per encounter. Hence, handling costs add up as diet breadth widens and broadening the diet breadth in the face of depletion of higher-ranked foods due to harvest pressure entails a decreased individual energetic ef ciency.

Hence, we can say that there has been a general historical trend toward lower individual energetic ef ciency (i.e. decreased net return rate) in human subsistence strategies, with a corresponding increase in spatial ef ciency, de ned as increased average total productivity per unit area of land (see further discussion of this issue below). Very broadly speaking, the adoption and culti-vation of domesticates can be seen at least in part as the culmination of this historical trend at the end of the Pleistocene.

A second broad implication of the diet breadth model is that the traditional concept of carrying capacity as a de nable limit or ceiling on growing population is far too simplistic. First of all, the model implies that, at any given time, there is likely to be a whole array of food items in the environment that may in fact be quite abundant, yet are not exploited because they are uneconomical to process. Second, it is clear that the relationship between population size and prey abundance is dynamic: as populations grow larger, prey abundance responds logistically-as prey abundance declines, ef ciency also declines and population growth is slowed. Finally, we can see that the commonly accepted de nition of carrying capacity as 'the upper limit of human population growth that can be achieved in a given habitat without eventual degradation of the resource base' (Hayden 1981: 412; discussed in Winterhalder et al. 1988: 322) is unrealistic: depletion is always occurring, and has an ongoing, dynamic relationship with forager population size. These implications are explored further below.

PUTTING THE DIET BREADTH MODEL IN A DYNAMIC, POPULATION ECOLOGICAL PERSPECTIVE

In its original form, the diet breadth model attempts to predict the optimal diet at a single moment in time-foragers will choose only those diet items from a ranked listed of potentially consumable prey that will result in the highest average net acquisition rate (NAR). The problem, as Gary Belovsky has succinctly put it, is that 'all food in the environment cannot be consumed without decreasing its recovery rate or decreasing the ability of the consumer to harvest it in the available feeding time' (1988: 330). Hence, both the optimal diet breadth and the NAR (i.e. ef ciency) of a population of foragers can be expected to change over time, with concomitant change in population growth rates and densities.

Exactly how do these variables change through time and in relation to each other and what kinds of environmental conditions-i.e. resource density and distribution-will affect how they change and how rapidly? As Winterhalder et al. (1988) have pointed out, this kind of population-resource dynamic is, as a general rule, empirically invisible to both ethnographers and archaeologists. It is invisible to ethnographers because they tend to observe human populations for relatively brief periods of time-what Winterhalder et al. refer to as the 'long ethnographic year'. Archaeologists, on the other hand, are in a position to observe such processes on a longer temporal scale, but, because of time averaging, they are typically unable to track population-resource dynamics at a scale ne enough to make real sense of it-one of the key aspects of this kind of process is that it is determined by day-to-day, individual decisions which have cumulative, recursive and, and in some cases, non-intuitive long-term consequences. Consequently, Winterhalder et al. argue that formal modeling techniques may be the most feasible way to get at population-resource dynamics over the long run. Such models are by necessity highly simplied, but, simple as they are, they have been shown to provide interesting and useful predictions about what we can expect.

There have been two major modeling efforts to put optimal foraging theory and long-term population processes into a dynamic, integrative framework with respect to human forager populations: the work of Gary Belovsky (1988) and that of Bruce Winterhalder et al. (1988). Both models attempt to integrate three dynamic processes that are key to the population-resource relationship: 1) the effects of resource acquisition and consumption on the human population growth rates, 2) the effects of changing prey densities on resource selection and 3) the effects of resource exploitation on population densities of prey. The models show how density of prey responds and changes as a function of their exploitation and how resource selection and growth rates of the forager population change as a function of resource density.

Belovsky's model combines a linear programming diet model that solves for the optimal mix of plant and animal foods with a spreadsheet-based population simulation that incor-porates nutrition intake required for somatic maintenance and the production of offspring. Belovsky imagines foraging environments that vary from low to high density with respect to primary harvestable productivity (PHP). Primary productivity refers to primary plant growth on the landscape upon which all herbivores and predators ultimately depend. Since a considerable amount of plant-food energy is typically locked up in a form not directly available to humans-i.e. cellulose-or is out of reach of humans-as in forest canopy-harvestable primary productivity refers to primary plant growth that is available for human forager consumption. In habitats with low PHP, animal prey are assumed to dominate the diet since what plant growth there is must be converted into usable form by animal prey. In habitats with intermediate PHP, animal and plant foods are present in more equally distributed moderate amounts and are more nearly balanced in the forager diet. With high PHP, both plant and animal prey are available in high quantities, but, since animal prey are assumed to be higher ranked due to their lower handling costs, animal prey initially dominate the diet.

When Belovsky combined the diet model with the population growth model, a population growth trajectory characterized by stable limit cycles was predicted at all levels of PHP. The severity (i.e. the amplitude, or ratio of peak to trough levels) of the limitcycles is determined by two factors: 1) the potential growth rate (predicted maximum level) of the forager population under a given level of PHP and 2) the predicted optimal diet breadth (i.e. in terms of relative dependence on animal vs. plant prey) at that level of PHP. On one hand, the higher the forager population growth rate, the more quickly it will deplete prey, initiating a new limit cycle. On the other hand, the broader the array of plants and animal prey that are available, the less precipitous the population decline phase will be, because, as higher-ranked animal prey are depleted, the population can fall back on less pro table plant foods. Hence, varying levels of PHP strongly affected the severity of the stable limit cycles.

Under low PHP, the forager population growth rate is low, and the limit cycles have lower amplitude because, when animal prey were depleted rst, there was little in the way of plant food to fall back on as diet breadth increased. Forager population levels are limited almost entirely by animal prey availability, and, as a result, foragers are unlikely to deplete animal prey populations entirely. Under intermediate PHP, animal prey are taken rst, and, as diet breadth broadens, the availability of lower-ranked plant foods dampens the severity of population decline: intermediate PHP produced the least severe limit cycles.

Under high PHP, limit cycles again increase in severity, because potential population growth is high, and, as higher-ranked animal foods are depleted, the high availability of fall-back plant foods keeps forager population levels high enough that they will continue Subsistence strategies and early human population history. Cycles are least severe at intermediate levels of PHP (400 gm/m^2). Average growth rates are about zero, even though growth phase rates can be relatively high; population density increases with PHP. Redrawn from Belovsky (1988).

to deplete animal prey, perhaps to the point of extinction. One interesting potential implication of this is that the adoption of domesticates would appear to be most likely in habitats with high PHP, as originally envisioned by Robert Braidwood.

DENSITY-INDEPENDENT FACTORS

In both the Belovsky and Winterhalder et al. models, stable limit cycles are due to density-dependent factors alone-population growth and decline and attendant prey depletion and regeneration. The addition of exogenous, density-independent factors such as climate-induced shortfalls would disrupt the pattern of stable limit cycles, depending on where in the cycle the shortfall occurs. For example, a brief (one or two season) shortfall in productivity coinciding with the beginning of a period of local population growth might have relatively little effect on a population. In contrast, even a brief shortfall coinciding with a growth peak or a population already into a decline phase, when density-dependent depletion has already started to occur, would likely have the effect of accelerating the decline and perhaps causing a serious crash or local population extinction.

For the purposes of the larger question explored in this discussion, that of changes in long-term human population growth rates, we can make the following observation at this juncture: in a perfect stable limit-cycle trajectory, the average long-term growth rate is zero, even though growth may be very high during one stage of the cycle and negative during another. Hence, any attempt to infer the 'typical' fertility behavior (or for that matter the mortality schedule) of foragers on the basis of long-term average growth rates alone is likely to be doomed to failure. Further, inferring typical fertility and mortality rates of foragers based on the few data points that are available from extant foragers is also problematic, due to potential sampling error over the whole cycle (Winterhalder et al. 1988: 320). On the other hand, if we can accept for a moment that human population trajectories are characterized in general by a saw-tooth, peak-and-trough pattern, it seems reasonable to suggest that broad, long-term increases in the average growth rate might be more realistically explained in terms of general factors that affect the average maximum that can be attained during a growth phase relative to the average minimum that occurs at the end of a decline or crash.

POPULATION DYNAMICS AND THE ADOPTION OF DOMESTICATES

The above models are designed speci cally for forager populations. What kinds of changes might we expect with the adoption of domesticates? Winterhalder et al. (1988: 323) suggest that, among foragers, depletion resulting from intensi cation of foraging activity both decreases individual net acquisition rate and diminishes total yield, inevitably reducing the sustainable human population of an area. Under domestication, intensi cation of production may reduce individual ef ciency, but, by de nition, increases total yields per unit area of land, which in turn increases the sustainable human population. Further, the resource base under domestication should rarely be subject to the rates of depletion that would occur under populations ultimately dependent on primary production in a natural environment. Hence, although shortfalls and population declines may continue under domestication, population levels among domesticators may not be subject to the same precipitous declines predicted by population dynamics models for foragers. This speci c aspect of domestication may in fact be one of its most critical distinguishing characteristics as a subsistence pattern.

There are at least two additional reasons why the adoption of domesticates might be expected to dampen the amplitude of growth and crash phases, and to push long-term average growth rates to a higher level. First, there is good reason to believe that the combination of high total productivity and storability of domesticates provide a buffer against temporary density-independent shortfalls caused by inter-annual variation in precipitation, length of growing season and the like. In contrast, when foragers undertake storage, it is primarily for getting through predictable seasonal resource shortages, not unpredictable inter-annual shortfalls.

Pennington (1992) studied the fertility and mortality of the !Kung in transition from a mobile to a more sedentary lifestyle that included some reliance on animal domesticates, and found that sedentism by itself did not signi cantly affect proximate determinants of fertility (length of reproductive span, length of birth interval-Harpending and Wandsnider (1982) had previously found similar results for !Kung living under nomadic and sedentary regimes). She did, however, nd that !Kung children living under sedentary regimes with animal domesticates had a 25 per cent higher chance of reaching maturity than those still living as mobile foragers, due to the availability of supplementary foods, particularly milk, during critical periods.

A second reason why the adoption of domesticates may reduce severity of population crashes is that increased dependence on crop agriculture and pastoralism have historically tended to promote increased differentiation in access to resources-that is, social ranking and strati cation.

From a population ecological point of view, unequal partitioning of resources within a population has, somewhat counterintuitively, the effect of increasing its stability over time (Lomnicki 1988: 20-34; Rogers 1992: 379-92). To see why, we can start by understanding unequal access to resources as the result of competition within a population over resources in nite supply. The character of resource competition can be thought of as varying along a continuum between two extremes that are termed scramble and contest competition (Boone 1992: 315-22). Scramble competition occurs when critical resources are economically undefendable, because they are either too unpredictably distributed in time or space or so thinly distributed on the landscape that they are not worth at any point putting up the costs of defense. Consequently, in a perfect scramble competition, all individuals in the population are said to have equal access to resources.

This in turn means that with each new individual that is added to the population, the feeding rate (and, ultimately, the tness) of all individuals in the population is reduced equally-individual resource availability is a function of group size. Now, in the most extreme case, one can imagine a growing population in scramble competition reaching a level such that the addition of just one new individual reduces the feeding rate of every-one else below what is necessary for somatic maintenance, and the entire population dies simultaneously of starvation-the proverbial straw that breaks the camel's back. This same result would occur in the instance of a sudden resource shortfall against a stationary population at the upper limit of its resource base.

Of course, this extreme situation rarely occurs, because there is always some individual variation in condition, energy requirements, ef ciency in acquiring resources and other factors that cause variation in instantaneous resource accessessentially, the true carrying capacity of a population will always vary from individual to individual. The fact remains, however, that populations in scramble competition are more susceptible to crashes caused by depletion of resources. Contest competition occurs over resources that are densely and predictably distributed on the landscape, such that they are economically defendable. In its most simplistic form, contest competition takes the form of a series pair-wise contests between individuals over resources, resulting in a ranked hierarchy based on differential access to defendable resources. When a shortfall occurs, individuals at the bottom of the hierarchy die rst, leaving the surviving population with a more favorable population to resource balance. In this case, crashes, when they occur, will be less severe than in the case of scramble competition. As a result, contest competition can be said to increase population stability (Rogers 1992: 386-7).

Generally speaking, foragers tend to exploit a resource base that more closely approximates a resource structure that promotes scramble competition, because resource density is lower, and the distribution of foraged resources on the landscape is less predictable than is the case with crop agriculture (Dyson-Hudson and Smith 1978). Hence, there is a sound theoretical basis for positing that forager populations are actually more, not less, susceptible to population crashes brought about by density-dependent and density-independent factors than domesticators. Populations that are less susceptible to crashes should have higher long-term growth rates even in the absence of any major increase in fertility. Each group starts with a population of 1000 and has a between-crash intrinsic growth rate of .007, a rate that has been calculated for the !Kung (Keckler 1997). When crashes occur, the foragers experience 25 per cent mortality, while among the farmers only 15 per cent of the population die. The simulation makes the simple point that two populations with the same growth rates during 'normal' periods can have very different long-term growth rates due to relatively small differences (10 per cent) in mortality during shortfalls.

DOES DOMESTICATION MEAN HIGHER FERTILITY?

In the above discussion, I suggested that higher average growth rates among agricultur-alists may result from the fact that they are in general less susceptible to population crashes due to altered resource depletion rates and changes in resource structure. Still, it is possible that general increases in average fertility may also be a contributing factor in increased growth rates under domestication, as has been commonly argued. I want to begin an evaluation of this issue with the observation that, over the past decade, closer analysis of foragers, horticulturalists and intensive agriculturalists from historical and ethnographic sources has revealed a surprising uniformity in fertility rates across all subsistence modes.

In 1988, Campbell and Wood published a cross-cultural compilation of total fertility rates (TFR) of seventy forager, horticulturalist, and intensive agricultural societies from the contemporary ethnographic record that showed there were no signicant differences in TFRs across subsistence practices. Hewlett (1991) published a similar analysis of forty mobile and sedentary foragers and pastoralists that indicated slightly higher fertility rates among pastoralists, although the difference was not signi cant. In 1993, Bentley et al. published an extensive critique and re-analysis of the Campbell and Wood study, presenting a new cross-cultural comparison of fty-seven forager, horticultural and intensive agricultural groups. Bentley et al. (1993a, 1993b) used a subset of the Campbell and Wood sample, excluding populations with high levels of sterility and also attempting to correct for the in ated sample size due to the inclusion of non-independent cases (ethnic groups that were closely related). These results showed no signi cant difference in TFRs between foragers and horticulturalists, and that intensive agriculturalists had signi cantly higher fertility rates than either of the other two groups (Fig. 3). All three subsistence categories had the same modal TFR of 6.0 live births per woman.

Bentley et al.'s nding that intensive agriculturalists have higher fertility than foragers, but that horticulturalists do not, strongly suggests that the speci c circumstances associated with the initial development of sedentism and higher total productivity associated with food production alone cannot account for the increase in growth rates thought to be associated with the adoption of domesticates. Further, all three subsistence groups show considerable variation in fertility, and subsistence mode by itself remains a poor predictor of fertility. As Pennington (1996) points out, knowing that a particular group has a mean TFR of 5.0 would not help very much in predicting that group's subsistence pattern. More recently, Sellen and Mace (1997: 886) carried out a phylogenetic analysis of the relationship between fertility and subsistence modes in sixty-nine cultures. Instead of Subsistence strategies and early human population history.

This essentially turned dependence on agriculture into a more or less continuous variable. Additionally, in an attempt to control for Galton's problem (the in uence of non-independence of populations in the samples due to historical relatedness), they constructed a global genetic phylogeny based on data on the distribution of human genetic polymorphisms collected by the Cavalli-Sforza team. First, within groups of phylogenetically related cultures, the best predictor of variation in fertility was variation in dependence on agriculture. Second, they used multiple regression analysis to test the association between dependence on agriculture and fertility by examining the variance in phylogenetically unbiased sets of contrasts rather than variance in the raw values (i.e. fertility rates and percentage dependence on agriculture) themselves.

Using a subset of the sample in which fertility estimates and estimates of dependence on agriculture were taken within twenty years of each other, the regression analysis showed that for each 10 per cent increase in dependence on agriculture, there was an increase of 0.4 births per woman. If we accept that intensive agriculturalists are in general more dependent on agriculture than horticulturalists and foragers, these ndings are generally consistent with Bentley et al.'s analysis.

Most explanations that attempt to account for a systematic relationship between fertility and levels of subsistence technology point to a speci c circumstance or characteristic associated with the technology itself and then posit how this circumstance affects specically the proximate determinants of fertility. For example, one widely held idea about why the fertility of mobile foragers should be lower is that mobility is incompatible with mothers caring for more than one non-ambulatory child at a time, and hence birth intervals among foragers are lengthened through various cultural practices, resulting in lower completed fertility (Sussman 1972). Elsewhere, the availability of weaning foods derived from crop plants has been cited as a mechanism leading to earlier weaning and shorter birth intervals following domestication (Buikstra et al. 1986).

Determinants of fertility fall into two general categories (Handwerker 1983: 6). The rst are proximate determinants (Wood 1990), which include such variables as the mean age at menarche, exposure to conception, lactational infecundity and the like. The second category-which Handwerker calls intermediate factors-consists of factors associated with political, economic and ecological conditions that ultimately determine the total time and energy budget that parents have available to produce and rear offspring. I want to suggest that, when we are talking about long-term patterns in fertility associated with subsistence, it may be more useful to begin by looking at broad patterns inherent in these intermediate variables-that is, those conditions associated with the overall energy budget of a household-and ask how subsistence technology relates to energy budgets and the trade-off between the time and energy devoted to child rearing relative to subsistence and other household maintenance activities.

Human families have a nite time and energy budget out of which various activities associated with survival, maintenance and reproduction must be funded. As noted above, time and energy expenditures are commonly subsumed under two categories: somatic effort, comprised of the time and energy spent in survival and maintenance; and reproductive effort, comprised of the time and energy expended in the production of offspring. Reproductive effort can further be divided into mating effort (nding mates) and parental effort (effort expended in producing and rearing offspring). At this point I am concerned mainly with fertility in already formed households, and will concentrate mainly on the trade-off between somatic effort (maintenance of the existing household) and parental effort. Hence, we will be interested in two general classes of economic activity: 1) the time and energy expended in acquiring or producing food, shelter and other resources to maintain the existing household and 2) the time and energy expended in provisioning children. The more time and energy that must be put into resource acquisition and processing to support the existing household, the less time and energy is left over for reproduction and rearing additional children. Hence, we might expect that the total number of children that a family can successfully raise should be directly affected by the net ef ciency of the family's food-and shelter-producing regime and how it relates to the total household budget.

In general terms, horticulture and agriculture both constitute forms of subsistence intensi cation. Subsistence intensi cation is a process by which the amount of time or energy required to produce and/or process food increases relative to the amount produced (Brook eld 1972). In other words, the labor cost per unit of food increases with subsistence intensi cation. Typically this occurs when items are added to the diet that require more time to harvest or process to render them edible and digestible, or because cultivation techniques are adopted that require more ground preparation and maintenance. In either case, the predicted consequence of subsistence intensi cation is decreased individual labor ef ciency; that is, an individual expends more energy for each calorie acquired or produced. The pay-off to intensi cation is that more calories per unit area of land become available, making it possible for more individuals to survive and reproduce on less land. In other words, subsistence intensi cation results in higher spatial ef ciency. This in turn results in higher local population densities.

There is, of course, no reason to think that all horticulturalists operate at a lower production ef ciency than all foragers, or that all intensive agriculturalists are less ef cient than all horticulturalists. In fact, it is clear that these subsistence categories overlap and represent only general tendencies. Variation in local ecology-particularly that which is related to the amount of biomass available for human consumption and the relative ease with which food production and processing can be accomplished-will cause variation in the level of production within a general subsistence regime.

Horticulturalists living in a rich temperate or tropical environment might well maintain production at a higher level of ef ciency than, say, mobile foragers living in an environment like the Great Basin of North America where animal biomass is relatively low and the labor cost to collect and process seeds and other plant foods is high. In fact, much of the variability and overlap in fertility rates between subsistence regimes indicated in Bentley et al.'s data may well be the result of variation in food acquisition ef ciency caused by differences in local ecology. However, we should expect that, in any particular region where intensi cation is taking place, there should be a general trend towards decreased ef ciency of food acquisition or production, and that any increase in fertility that might occur after the adoption of a more intensive technique must come from gains in handling or processing ef ciency, improvements in the productivity of cultigens and the like.

Two important implications ow from this perspective on the energetics of subsistence intensi cation and the trade-off between resource acquisition and reproduction presented above. First, if individual ef ciency (labor per unit of food) declines under intensi cation, holding other conditions equal, it seems reasonable to expect that fertility should actually decrease, not increase, with subsistence intensi cation, since parents have to work harder and longer to support themselves and their offspring. Second, in an area where subsistence intensi cation is taking place, we should expect local population density to increase as a result of the increased spatial ef ciency of food acquisition or production, regardless of whether the population growth rate increases or stays the same.

Hence, one expectation is that, with the increase in subsistence intensi cation and processing costs associated with domestication, fertility initially may actually decrease, even while local population densities increase because of increased spatial ef ciency of production (Hawkes and O'Connell 1992: 63). Bentley et al.'s (1993a, 1993b) data on the relative TFRs of foragers, horticulturalists and intensive agriculturalists, discussed above, would seem to support at least part of this proposition in the sense that horticulturalists on average have about the same or even a slightly lower TFR than foragers.

To summarize thus far, the only prediction that we can really make about the relationship between level of subsistence intensi cation and fertility is that, ceteris paribus, fertility ought to decline with increased subsistence intensi cation. The fact that it does not-at least with respect to intensive agriculture-then becomes an interesting question. How can parents, if they are working at a lower ef ciency, afford to support yet more children than before? I discuss two possibilities below. One possibility is that the additional effort comes from males. There is abundant evidence that under intensive agriculture the relative contribution of males to food production increases as the total household workload increases (Ember 1983, although Ember views this in terms of a relative decrease in the participation of females).

A fairly robust de ning feature of horticultural societies is that the bulk of the food energy consumed is produced by women. Under these circumstances, the effect of male contribution to subsistence on the survival of offspring is low. Hence, in the trade-off between parental effort and mating effort, males in horticultural societies should invest more in mating effort and in malemale competition over access to females (Lancaster and Kaplan 1992; Lancaster 1997). These appear to be the main determinants of the matrifocal, matri-lineal character of the majority of horticultural societies. Under agricultural intensi cation, the workload necessary to provide suf cient food to raise offspring increases to the point where women cannot do all the work. At this point, the trade-off for males between parental effort and mating effort shifts back in favor of parental investment-male PI becomes critical to survival of offspring.

The result is a shift to patrifocal, patrilineal family systems, where males contribute a signi cant proportion of labor to the household budget. Another possibility is that the net contribution of children to the household budget increases under intensive agriculture. In the few quantitative studies of production and consumption over the life cycle that have been carried out on forager and horticulturalist groups, the available data have shown that children do not produce more food than they consume until about the age of 20 (for Piro and Machinguenga horticulturalists, Kaplan 1996: g. 4; for Ache foragers, Kaplan 1996: g. 4; Kaplan et al. 2000: g. 3), a time when they begin to marry and produce children of their own. Kaplan (1996) argues that the reason net production in children occurs so late in foragers is that the skills required to carry out ef cient hunting and collecting activities take this long to acquire, a factor which, he argues, explains the evolution of the long period of juvenile dependence in humans (see also Kaplan et al. 2000).

In contrast, Kramer (in press) carried out a similar study of production and consumption over the life cycle among traditional Maya agriculturalists, where she found that girls achieved net surplus production by age 12 and boys by age 17. Since they do not marry and start new households until the age of about 19 or 20, there is a period of up to about six years when children are net contributors to the household fund. In an analysis of the total amount of labor required to maintain a typical Maya household (measured in hours of labor per week), Kramer also found that parents were not able to produce all the required work alone. Hence, she argues that the labor contribution of older children actually underwrites the production of additional younger children in the Maya household by reducing the workload on their mothers (mean fertility in her sample was 7.0).

In summary, the picture that is beginning to develop is that the speci c circumstances surrounding the adoption of domesticates and attendant sedentism are not themselves suf cient to explain observed increases in fertility, since available evidence suggests that signi cant increases in fertility do not occur until dependence on agriculture is already well established. Rather, changes in the household energy budget resulting from reorganization of the relative contribution of men, women and children may turn out to be a more promising explanation for increases in fertility that occur with the adoption of agriculture.

CAN CATASTROPHIC MORTALITY EXPLAIN NEAR ZERO GROWTH?

Our present knowledge of the relation between subsistence mode and fertility indicates that increases in fertility under agriculture may at least partly explain the apparent increase in growth rates under domestication, although such increases may not have become a factor until agriculture was relatively well established. Since the data on subsistence mode and fertility reported in the above discussion are essentially synchronic in nature, they cannot be used, strictly speaking, to test directly the diachronic proposition that increased dependence on domesticates causes increased fertility (Sellen and Mace 1997: 287). However, the data do indicate that foragers are capable of fertility levels as high as agriculturalists, and no forager group has a fertility level low enough to explain the near-zero growth that must have been in place through most of the Pleistocene. We are left with the possibility that infrequent catastrophic mortality interspersed with periods of relatively rapid growth explains this pattern. Unfortunately, the empirical data needed to test such a proposition are even less forthcoming than those relating to subsistence and fertility. The following observations, however, can be made.

Classical models of logistic population growth assume that animal populations grow until they reach a stable state at or near K. Recent empirical work in population ecology, however, has shown that many animal species, including large mammals, are characterized by periods of growth interrupted by relatively infrequent crashes brought about by starvation, disease and other environmental factors that may be, to varying extents, density independent (Dunbar 1987: 77; Mangel and Tier 1993; Young 1993).

Young (1993) surveyed a series of eighty documented natural die-offs in large mammals, including herbivores, carnivores and primates, which were characterized by peak-to-trough reductions in population of at least 25 per cent. Within the sample, Young found that the modal degree of population reduction during crashes fell in the range of 70 per cent to 90 per cent, although he suggested that this might be partly due to underreporting of die-offs of lesser severity. There was also a pronounced paucity of die-offs involving reductions of greater than 90 per cent, which Young suggested may be due to the fact that natural selection tends to eliminate populations susceptible to total extinction (ibid.: . 414). Interestingly, all three classes of mammals (herbivores, carnivores and primates) were equally represented in the 70-90 per cent range, although carnivore die-offs were more likely to be caused by disease, while herbivore and primate die-offs were more likely to be caused by habitat decline and starvation.

The question then arises as to what extent human populations have been subject to periodic die-offs in their evolutionary history.

Several lines of indirect evidence lead to the conclusion that infrequent, but severe population crashes may have been a chronic condition of human population history.

First of all, as pointed out at the beginning of this article, it is difficult to reconcile the high rates of growth (and the potentially high reproductive capacity) observed in modern human populations, including modern foragers, with the extremely low overall rates of growth estimated throughout most of human history unless one argues that periods of relatively high growth were counter-balanced by periodic population crashes.

Hassan (1981: 254), using the geometric growth equation in connection with plausible estimates of population sizes at different points in the European and Middle Eastern prehistoric sequence, has estimated that at most overall growth rates up until the Neolithic transition ranged between .00007 and .011 per cent per year. Yet even the modern !Kung, who live in a relatively marginal foraging environment, have a calculated growth rate of .7 per cent (r = .007) (Keckler 1997), while a tropical foraging group, the Ache of Paraguay, sustained a pre-contact growth rate of 2.5 per cent (r = .025) (Hill and Hurtado 1996: 258).

To see the problem more clearly, we can take the !Kung growth figure of r = .007 and apply it to a starting population of 10,000: such a population would grow to the present population of the earth in about 1900 years.

Based on the fertility data collected by Campbell and Wood (1988) and Bentley et al. (1993a, 1993b), Hill and Hurtado observe that, on average, 'the maximum interbirth interval that can be attained in poorly nourished, non-contracepting populations with late weaning is about 48 months... [this] implies six live births if reproduction begins at age twenty and ends at forty years of age' (1996: 471).

Since most studies of such populations indicate a mortality rate of about 50 per cent, this implies growth rates far above what could have been possible during much of human history. Hill and Hurtado argue that, with a an average TFR of 6.0, juvenile mortality would have to have been close to 67 per cent to produce near zero growth, yet no traditional population with such high rates has ever been observed.

Another line of empirical evidence for catastrophic mortality in more recent prehistory comes from mortality pro les of prehistoric skeletal populations (Keckler 1997). Such populations often exhibit a high proportion of primeaged adult deaths, resulting in a attened mortality pro le, compared to the usual U-or J-shaped pro le (with higher mortality among the young and old) that typically characterizes mammal populations.

These attened mortality pro les do not easily t any known model life table. Keckler examined four prehistoric skeletal populations: a combined sample of Neandertals, and three forager-horticultural series from North America-Libben, Dickson Mounds 1 and Oneota. In a series of simulations using life-table characteristics of two hunter-gatherer populations, the !Kung and the Ache, as well as the Coale and Demeny West Model 1 life tables, Keckler showed that, by adding random population crashes involving reductions ranging from 10 per cent to 54 per cent at a mean rate of one every thirty years, he could generally produce both the near-zero overall population growth rates estimated by Hassan as well as the attened mortality pro les observed in four skeletal populations.

Keckler concluded that human population growth in the past is likely to have been characterized by what he refers to as a biphasic pattern: relatively long periods of growth counterbalanced by short intervals of catastrophic decline.

The main problem with this study is one of equi nality. Lower archaeological recovery rates for juveniles and older individuals could also produce higher proportions of primeaged adults in cemetery populations. However, Keckler's alternative interpretation of skeletal mortality patterns with high rates of adult deaths is tantalizing, and analysis of more large skeletal series with an eye toward ltering out the effects of differential recovery rates may resolve this issue.

One promising avenue of empirical investigation of this problem comes from analysis of genetic variability in modern human populations. Recent analyses of variation in modern human mtDNA have shown that episodes of population growth and decline leave characteristic signatures in the nucleotide site differences between pairs of individuals (Harpending and Rogers 1993). Harpending et al. (1993) presented an analysis that indicates that the modern human population arose from a population expansion that occurred between 30,000 and 80,000 years ago from a prior effective population size of only 1000 to 10,000 females. Ambrose (1998) has argued that one or more bottlenecks occurred in the late Pleistocene due to climatic downturns.

One event in particular appears to have been caused by the eruption of Mt Toba, Sumatra, approximately 70,000 years ago, which may have decimated human populations outside tropical refugia. While these proposed bottlenecks are clearly on a much larger scale than the kind of local or regional crashes proposed in this article, they may indicate that population crashes have played a signicant role in shaping the structure of modern human populations.

CONCLUSIONS

In the above discussion, I have made a number of claims regarding the nature of early human population history, nearly all of which have been made previously by a wide array of investigators in the eld of evolutionary anthropology, very few of which have been widely recognized or accepted in the eld of anthropological archaeology.

The revised picture I have attempted to put together is by necessity very broad and tentative in nature. However, several preliminary conclusions can be drawn.

First, the idea that carrying capacity represents a static ceiling around which human populations maintain an equilibrium has been outmoded for some time. It has begun to be replaced with more dynamic view of the relationship between population density, prey density, individual energetic efficiency and reproduction rates.

Second, the fertility rates of traditional populations across all subsistence modes vary considerably around the same mode and nearly the same mean; the continued search for ways in which the specific circumstances of subsistence strategies tweak the proximate determinants of fertility does not seem a very profitable avenue of investigation for explaining broad changes in growth rates in human history. Rather, a closer look at the ways in which individual energetic efficiency, resource depletion rates and resource structure vary generally across subsistence modes and affect population growth rates and densities appears to be in order.

Finally, the idea that infrequent, but serious population crashes have played a substantial role in shaping past human population trajectories deserves a closer look. Population crashes are rare events viewed from the perspective of the human lifespan-if they were not, there would be no one alive today to muse on the human condition. The fact that they are so infrequent appears to have led to the perception that they are abnormal or atypical events.

9

Building the Contextual Milieu: An Approach to Data Collection and Interpretation

In this chapter an integrated approach to archaeological data collection and interpretation is offered, based upon the fundamental idea that all archaeological materials were deposited in a complex, interactive, contextual milieu, comprised of the physical, biological and cultural aspects of a past environment. Further, it is apparent that these useful but arbitrary divisions of the contextual milieu were intimately interconnected and, while they can be readily separated for analytical purposes, this interconnection must be recognized to fully understand the context in which past societies operated.

A brief overview of recent theoretical positions is offered, followed by a discussion of the major aspects of the physical, biological and cultural environment and their significance for human populations, followed by a discussion of the classes of evidence upon which our analyses relied. This paper offers a contextual interpretive framework using examples from the SCAPE Project to illustrate the utility of an applied theoretical approach to the collection and subsequent interpretation of archaeological data.

Over the past five years of field work in the SCAPE Project, a working relationship was developed between the various disciplinary specialists that led to a cooperative focus on the development of reconstructions, and the modeling of past environments, including the physical, biological and cultural parameters, that placed a variety of challenges and changing opportunities before human groups at different times and places in the near and distant past. This interdisciplinary orientation has been driven by the need to critically address the interpretative context of archaeological sites, their contents and their larger landscape associations. This synergy and knowledge sharing has been subsumed under the concept of developing a contextual milieu, in which past societies lived, interacted with other societies, and with their immediate environments. In establishing a relevant contextual milieu, it is our intention to develop environmental settings that transcend the usual concept of a biophysical environment that includes, in addition, the cultural environment of the time. In this way it is believed that a given archaeological culture and its members can be more fully understood.

In this volume the idea of building a contextual milieu is presented. This is primarily an applied approach that places the human group as the focus of interest. The human group is situated in a tripartite context that draws upon the data from the physical, biological and cultural aspects of the environment to build a contextual milieu in which these groups, at different times and places, conducted their affairs. The milieu is viewed as being dynamic and interactive with emphasis being placed upon the role of human culture in mediating between the group and the challenges and changing opportunities their tripartite environmental context presented.

BUILDING ON THE WORK OF OTHERS

The formulation and development of archaeological method and theory has taken place over a significantly long period of time. Beginning with the musings of antiquarians, and through speculative historical reconstructions, an outline of the progression of human groups across time and space was fabricated. Necessarily, most of this early work has been amended or repudiated with the advent of better methods of data recovery and an understanding that the range of experience of past human groups varied as widely as the experience of those in the present. It can be anticipated that this intellectual progression will continue for the foreseeable future. A number of useful approaches to data recovery and interpretation exist within the discipline at the present time and a brief review of the literature over the past four decades traces the evolution of the initial proposals into configurations that are more comprehensive, efficient, accurate, and more effective in their predictive capabilities.

Among these are Landscape Archaeology, which blended geoscience with archaeological interpretation to understand the evolution of landscapes, Contextual Archaeology that began with a call for a reconstruction of environments including physical, biological and cultural contexts (Butzer 1982), and Historical Ecology that postulates a dialectic between the natural environment and culture (Balée 1998; Crumley 1993). All of these approaches have evolved through time to broaden their usefulness or more clearly define their methods and goals. Landscape Archaeology has significantly broadened its scope to include a stronger role for the cultural component and Contextual Archaeology has progressively shifted towards becoming an interpretive paradigm coming into play once the data has been recovered. Historical Ecology, presuming somewhat detailed data sets for both the natural and cultural parameters for the times and places under investigation, is most useful in more recent contexts.

The foundations for modern archaeological practice were laid out in the late 19th century, in Europe by wealthy individuals such as General Pitt-Rivers who excavated extensively at Cranborne Chase in the late 19th century. These individuals laid the foundations for a scientific approach to archaeological excavation. Similarly, in the New World, a more rigorous descriptive phase developed, based upon careful excavation of archaeological sites (Ford and Webb 1956; Kidder 1924; Moore 1905; Nelson 1914).

Probably the most influential theorist in the rise of a culture/environmental approach to archaeological interpretation was Julian Steward. Working primarily in the "Great Basin", he came to the conclusion that the dynamic between environment and culture led to similar socio-cultural configurations emerging repeatedly in similar environmental contexts. Steward (1955) reached three basic assumptions concerning the evolutionary development of societies: 1) similar adaptations may be found in different cultures when their environments are similar, 2) no culture has ever achieved an adaptation to its environment which has remained unchanged over any length of time and 3) differences and changes during periods of cultural development in an area can either add to social complexity or result in completely new cultural patterns.

Steward (1955:93) noted, "... that functionally interrelated economic, social, political, religious and military patterns, as well as technological and esthetic features, have become the basis for development taxonomies." He believed that these societal characteristics formed the cultural core of a given society. People who shared a cultural core interacted with all aspects of their environment, enabling a given society to successfully respond to changes and possibilities that they faced. In this way cultural types could be understood in relation to their particular environments and their evolution history.

This Cultural Ecology paradigm was succeeded by an empiricist processual agenda that called for hypothesis testing and a much more comprehensive understanding of the technological developments and the processes, that characterized individual societies in the near and more distant past. The call for change was first clearly articulated by Walter W. Taylor in his monograph, A Study of Archaeology (Taylor 1948).

He advocated a change in direction that he called the Conjunctive Approach and, by connecting artifacts with their cultural contexts, would provide an empirical base for understanding how past societies operated. The goal was to understand how the ordinary people in the society lived and made their contributions, rather than focusing almost exclusively on the roles of the society elites. His criticism of the work of prominent archaeologists of the day was not well received at the time. A decade and a half later Binford, in a series of articles (Binford 1962, 1964, 1965, 1972) renewed the call by Taylor for a more scientific approach to archaeological interpretation and he introduced the term Processual Archaeology. The timing was more conducive to acceptance of a scientific paradigm and Binford, and his students and colleagues, enjoyed wide acceptance within North American archaeology. Others who shared these ideas (e.g. Thomas 1973; Schiffer and Rathje 1973) produced works of lasting significance that contributed to a major paradigm shift at this time.

Thomas (1973), in his Reese River Valley study firmly linked Cultural Ecology to the new Processual Archaeology paradigm. Thomas undertook the empirical testing of Stewards (1938) family based Shoshonean seasonal round subsistence pattern. The data for Stewards reconstruction was largely based upon information from Shoshonean informants, as well as a study of the environment, which had been seriously impacted by European settlers. In reviewing Stewards work, Thomas (1973:156) states that, "The seasonal round-taproot of the entire Shoshonean cultural adaptation-is simply not a fact; it is a theory." Thomas began his testing of this theory by extracting data from the ethnographic work of Steward (1938) to build a simulation model of the Shoshonean seasonal round, producing a flow chart of Shoshonean subsistence (Thomas 1973:159). Thomas then produced 130 hypotheses to be tested empirically (Thomas 1971, 1972). Of these, 75% were confirmed. Thomas (1973:172 concludes, "... it is probable that the Shoshonean subsistence system operated in the prehistoric Reese River Valley". This processual approach to a largely ethnographic, theoretical construct was well received and added immeasurably to the reputation of processual archaeology. It also linked the importance of environmental studies to verifying cultural interpretations.

This "new Archaeology", used an explicitly scientific approach, following the ideas of Carl Hemple, a logical empiricist philosopher of science. Hemple believed that scientific explanations do not depend upon the individual proposing them. "They depend instead upon laws that describe the phenomena under investigation and on establishing the relevant initial conditions for the case at hand" (O'Brien et al. 2005:92). These observable phenomena are the materials made, used and discarded by past societies. It was believed that these data could serve as proxy measures for their makers' behaviors and their social constructs. These insights would serve as a basis for discovery of laws explaining past human behavior.

Environmental Archaeology This trend towards a more scientific archaeology was accompanied by a refocusing on the environment-specifically its physical and biological aspects. An early proponent of this approach was Karl Butzer (1971). In his forward to Environment and Archaeology Butzer (1971:xi) stated that, "In its present form this book is specifically intended as a text in Pleistocene geography for coursework in both Pleistocene geology and geomorphology, and prehistoric archaeology. Possibly the book may also encourage closer relations between the natural scientists and archaeologists-in the field, the laboratory and the university."

The term environmental archaeology has been used in the literature in a number of cases to present the idea that archaeological materials are situated in an environmental context. Usually, this environment was seen as being limited to the "natural environment" (Jochim 1976, 1981; Winterhalder 1981). Humans are seen as typically responding to environmental opportunities using a reasoned, optimal decision making strategy (Winterhalder 1981).

Butzer (1982) returned to the idea of environmental archaeology in another monograph, Archaeology as Human Ecology, and advocated a much more comprehensive approach to the idea of environment and advocated that a contextual approach be taken. In his view, "The basic ingredients of archaeology are artifacts and their context, ranging from food residues to sediment and landscape matrix. The term context means many things to many people, but the term is derived from the Latin verb contexere, "to weave together" or "to connect." For archaeology, context implies a four-dimensional spatial-temporal matrix that comprises both a cultural environment and a non-cultural environment and that can be applied to a single artifact or a constellation of sites (Butzer 1982:4)".

Butzer (1982:12) disputed the idea that archaeology can and should be subsumed into the discipline of Anthropology. He argues that while Anthropology can provide valuable socio-cultural models for archaeological interpretation, the discipline is similarly dependant upon the geo-sciences for understanding the events of the near and distant past. This being the case, a broadly based interdisciplinary approach contributing to a contextual archaeology paradigm is essential. He states that:

"I am therefore arguing for a contextual archaeology rather than an anthropological archaeology. My plea is for deliberate exploration and development of an approach that will transcend the traditional preoccupation with artifacts and with sites in isolation, to arrive at a realistic appreciation of the environmental matrix and its potential spatial, economic, and social interactions with the subsistence-settlement system. The human ecosystem so defined will open up truly ecological vistas that have been largely neglected. This contextual approach, heavily dependent on archaeobotany, zoo-archaeology, geo-archaeology, and spatial archaeology, is new not in its components but by virtue of its integrated general goal of understanding the human ecosystem" (Butzer 1982:12).

The contextual archaeology advocated by Butzer (1982) remained heavily committed to the development of the geo-science and bioscience aspects of the environment (Butzer 1982:Part II). In Part III, Butzer focuses primarily on human settlement patterns with a brief nod towards culture as a dynamic adaptive system, interacting with the natural environment (Butzer 1982:281-282). However, he emphasizes the role of the natural environment as the determining factor and arbiter of successful adaptive strategies, with historic practice and culturally mediated desires as minor determinates of a successful strategy (Butzer 1982:284-285). There remains, of course, the question of whether success is measured by the emic perceptions of participants in a given culture or the supposedly dispassionate etic observations of contemporary scholars. While Butzer envisioned the value of a contextual archaeology that embraces the physical, biological and cultural aspects of environment, his focus was on landscape and the impact of people on that landscape, rather than attempting to deal with their aspirations and their socioeconomic and sociopolitical goals, that may also be reflected in the archaeological record.

Building upon the work of Steward (1938,1955), Winterhalder (1981), Butzer (1982) and many others, the environmental archaeology paradigm continues to evolve, as authors amend and revise the practices and goals of this approach. Dincauze (2000) offers a detailed manual for the implementation of an interdisciplinary approach to environmental archaeology. This work constitutes an authoritative guide to a variety of paleoenvironmental, geoarchaeological and archaeological methods, ranging from dating techniques, to climatic reconstruction, geomorphology, sedimentology, soil science, paleobotany, and zooarchaeology. Dincauze envisions this work as being a foundation for Human ecology and states that,

"The following chapters will amply demonstrate that an ecological perspective on human physical and social history and evolution is by no means determinative. People are the proximate causes of change in their societies; their environments reflect, amplify, or dampen change, and return changed conditions to the instigators of change, requiring new adaptations." (Dincauze 2000:18).

However, notwithstanding the above disclaimer, this excellent work is clearly focused upon the biophysical environment with human groups and their constituent members playing a largely dependent and reactive role within their biophysical environment. This work, and that of others, sets the stage for the emergence of landscape archaeology.

Landscape Archaeology More recently, many archaeologists and geographers have adopted Landscape Archaeology, as a means of situating people in the broad context of landscape. This has proven to be a useful approach to problems that are primarily concerned with landscape evolution and settlement patterns. The antecedents of landscape archaeology extend to the Viru Valley Project that Willey (1953) conducted under the theoretical framework of settlement pattern analysis. The project incorporated a multidisciplinary approach to locating resources and settlements in the region under study. Similarly, the work of Graham Clark (1954) reconstructed landscape and subsistence potential at Star Carr, a Mesolithic site in Yorkshire.

Landscape archaeology has also been used for smaller projects such as the interpretation of the 18th century Georgian garden of William Paca in Annapolis (Leone1984). Similarly, Crumley (1993:6) has emphasized the dynamic relationship between landscape and humans in her description of landscape archaeology as, "... the material manifestation of the relationship between humans and their environments." Landscape Archaeology, like other approaches to the collection and interpretation of archaeological data, continues to evolve. Wandsnider (1992:290) sees Landscape Archeology primarily as a means for understanding settlement patterns, following the goals first put into practice by Willey (1953). However, the focus on the landscape, with humans as vectors that impact on landscape evolution, remains. As an illustration, this widely accepted understanding is reflected in a recent definition of Landscape archaeology provided by the Archaeology Web Page:

"**Definition:** Landscape archaeology studies the way people of the past shaped the land around them, consciously or unconsciously. In the most extreme, shaping the environment might mean moving earth, but it also involves gardens and pathways and buildings. Landscape archaeology could almost be described as a more holistic form of archaeological study than site-specific archaeology; and it often includes GIS mapping of the "anthropogenic" or human-built environment."

As the naming of the paradigm implies, this current definition of Landscape Archaeology clearly places the focus of Landscape Archaeology on the human agency in landscape evolution and the role that human groups have played in modifying the landscape to varying degrees. These modifications range from trail systems and the disturbance that radiates from these networks, to anthropogenic burning, to the land development schemes found in ancient and more recent times. Although sometimes implicit, an understanding of the historic, socio-political and socio-economic dynamics of these groups is not central in this approach.

Historical Ecology Historical Ecology has been widely accepted as a theoretical guideline by researchers working in times and places where ethnohistoric, historic, or traditional oral narratives are available. Historical Ecology is a fairly new field of study that takes a human/nature dialectical approach to the history of landscapes, cultures, and regions. Where Historical Ecology differs from Landscape Archaeology is the recognition and predominance of anthropocentric forces upon the landscape. One of the pioneers of Historical Ecology, Carole Crumley (1993:4-5), notes that Historical Ecology traces the ongoing dialectical relations between human acts and acts of nature, made manifest in landscape. Her belief is that the physical evidence of these activities provides the scholar a totality for study and analysis that is multiscalar, diachronic, and holistic. In contrast, in a recent book, Advances in Historical Ecology (Balée 1998), there is little agreement on a definition of what Historical Ecology really is, and the lead article by Whitehead (1998:31) argues that Historical Ecology offers nothing not already offered by Historical Anthropology and is therefore little more than an interesting methodology. Like many others, this paradigm continues to evolve.

Contextual Archaeology: The contextual archaeology approach advocated by Butzer (1982), has been extensively modified by Ian Hodder. Hodder has had a long-term interest in the idea of context and the interpretation of archaeological materials (Hodder 1986, 1987, 1989; Hodder and Lane 1982). In the recent archaeological literature, Hodder (Hodder and Hutson 2003) has modified the more applied approach advocated by Butzer (1982), into an interpretive paradigm that relies heavily on the manipulation of data to achieve an understanding of meaning in prehistoric communities.

In this approach, the traditional analysis and interpretation, routinely conducted by archaeologists, is subjected to hermeneutic analysis to achieve understanding (O'Brien et al. 2005:233). The focus for Hodder is the context, in the broadest sense, of the object. Text, he believes, is created as a metaphor for material culture through analysis and the interpretive process (Hodder and Hutson 2003:169).

This metaphoric text requires hermeneutic analysis to achieve understanding (Hodder and Hutson 2003:204). In Hodder's view, as archaeologists we move back and forth between the portion of the archaeological record under study and our understandings of natural and cultural formation processes until the two are aligned (O'Brien et al. 2005:233). If this reconstruction makes sense, then we believe that we know what or who created that record. This theoretical discourse, with its plethora of cautions, is stimulating reading and contributes to our understanding of the ideas that underlie the interpretations and the means of understanding the archaeological record. However, it provides little that can be directly embedded in archaeological practice-particularly in the field-data collection phase. This is a general problem with theoretical debates in archaeology and it is recognized by Hodder (2002:87) that there is a "... trend that can be recognized as archaeologists face a widening debate in a new global context (that) concerns the embedding of theory in practice".

The basic goals and practices employed in building a contextual milieu agree more closely with the earlier definition of contextual archaeology formulated by Butzer (1982) than with the theoretical perspectives of Hodder (e.g. Hodder and Hutson 2003). The building of a contextual milieu provides a way to embed a wide range of theoretical and methodological constructs into archaeological practice. This embedding begins with the formulation of the research proposal and the selection of a research team, and continues through the collection and interpretation of data, to the joint publication of research results. Further, the focus in building a contextual milieu is clearly directed towards providing a context for the activities and practices of particular groups of people at specified times and places, in the near and distant past. The tripartite environmental reconstructions provide a context for understanding the variables affecting the human groups being studied. An understanding of these groups, placed in time and space, is the focus of primary research interest.

AN APPLIED APPROACH TO BUILDING A CONTEXTUAL MILIEU

Following upon the brief review of the development of several paradigms offered above, an applied methodological and theoretical approach to studying human groups within a tripartite context is provided. The principal difference between this approach and landscape archaeology is that the focus of the exercise is upon human groups conducting their affairs within as fully developed a context as possible, rather than seeing these groups primarily as factors influencing the development-for good or ill-of the landscape that they move upon.

It has been widely recognized for many years that the physical and biological environments impose certain limitations on the activities and subsistence possibilities for human groups (Jochim 1976, 1981; Keene 1981; Winterhalder and Smith 1981). It is also widely recognized that culture, a uniquely human adaptation, mediates between people and their environment, ameliorating these constraints through technological and socio-economic innovation and through a retained knowledge of traditional lore and practice (Nicholson 1987:59-60).

Since so much of human cognition, adaptation and even social innovation, are conditioned by learned behaviour and traditional values, human decision making is normally cautious and based upon past experience and reflects efforts to satisfy immediate needs and priorities. When previously reliable resources fail or become scarce, the innovation or extinction options may require substantial and rapid adjustment to longstanding practices. While the consequences may be far reaching in terms of changes in adaptive strategies and socio-economic organization, humans can act and react with rapidity and a new flexibility that might ideally be measured in days or weeks.

With this realization in mind, researchers immediately confront a major challenge in that archaeological control over the temporal variable is, at best, coarse-grained. Even the highest resolution radiocarbon dating can seldom achieve better than decadal or century level temporal resolution. Similar imprecision confronts our efforts to understand archaeological deposits in a spatial context. While archaeological excavation might achieve considerable precision in three-dimensional recording of artifact distribution, it is readily apparent that aeolian and fluvial processes, pedogenesis, bioturbation, decay, and cultural transformation processes are at work from the moment that the material culture is deposited until it is collected (Clark and Keitzke 1967; Lyman 1994).

The spatial context of human activity can also be comparatively small-scale and localized. Archaeologists often seek to comprehend human action at the individual or household scale, within a spatially circumscribed village, or within its immediately surrounding 'catchment area'. However, the contextualization of human settlement and land use also takes place within a context that can encompass much larger-scale biotic, physiographic and climatic zones (Wandsnider 1998). Similarly, the social context of human activity varies from family level nomadic seasonal rounds (Steward 1938), through band and macroband organization, to much more complexly organized village or state levels of social organization. Alliances, formal and informal, also play important roles in the redistribution of goods and access to resources at all levels of social organization.

As the interpretative capacity of environmental archaeology and geoarchaeology has developed, the pragmatic limits of traditional archaeological inference have been increasingly exposed to critical scrutiny. There has been a tacit recognition that an explicitly interdisciplinary research perspective is required to maximize data recovery and the interpretative resolution of the analyses (Braidwood and Howe 1960; Byers and MacNeish 1967-1976; Butzer 1971, 1982). To more fully understand past human cultures, the physical, biological and cultural aspects of the tripartite context in which a given cultural group functioned must be reconstructed as accurately as possible.

Axiomatic Guidelines In order to proceed with the building of an explicitly interdisciplinary contextual milieu, it is useful to present axioms/postulates that can serve as guiding principles. It is important, from the perspective of clarity and cohesiveness, that such guiding principles be clearly articulated, so that there is a consistent approach to archaeological data collection, analysis and interpretation. This will enable the creation of an understanding among team members of the nature of the platform from which protocols for data collection and the articulation of shared goals can be derived.

ENVIRONMENTAL PARAMETERS, METHODS AND CONTEXTUAL ANALYSES

Increasingly refined methods of archaeological site investigation and more rigorous interpretative inference have become commonplace since the 1960s, particularly with the rapid maturation of geoarchaeological and archaeometric field and laboratory methodologies. The SCAPE project has taken advantage of this growing analytic sophistication with the integration of several investigative approaches by our team. Perhaps the most fruitful aspect of this integration is the sustained field and lab collaboration between archaeologists, ethnohistorians, sedimentologists, geoarchaeologists, physical geographers, palynologists and other paleoecologists (Running et al. this volume). The sharing of ideas and healthy debate over interpretation has contributed to a working environment of collective learning that has significantly furthered the professional development of the co-investigators as well as several hundred undergraduate and graduate students. At the methodological level of site investigation, systematic fine screening, routine sediment sample collection, deep sediment coring, three dimensional artifact documentation, integration of total station and differentially corrected GPS technology, widespread use of digital photography, GIS supported spatial data collection, analysis and visualization, and geophysical remote sensing, represent only some of these innovative collaborations. Beginning circa 1990, the use of GIS technology in the modeling of settlement patterns has increased in importance (Ebert 2004; Hamilton 2000).

More significantly, it has been possible to avoid the mere appropriation of methods and concepts from other disciplines and their uncritical integration into the archaeological enquiry (see Dincauze 2000:21). With a team approach, individual researchers do not work in intellectual isolation. With regular field interaction between research specialists, team members could propose multiple working hypotheses (often while in the field), and have them subjected to critical scrutiny and refinement, alternative interpretation and iterative debate. This debate ultimately led to a better understanding of all of the components in the complex milieus under investigation and culminated in more influential co-authored publications. Through the close integration of scholars in cognate disciplines, the depth and sophistication of interpretations were considerably enhanced. This is particularly the case in terms of more credible and fine-grained climatic and ecological contextualization of individual sites and localities by specialists, rather than through the sometimes novice efforts of the archaeologists. Similarly, anthropological perspectives brought to the collective analysis, have served to temper and deepen sometimes overly deterministic interpretations of human adaptation that might derive from over-dependence upon natural science and from theoretical evolutionary and ecological perspectives.

Physical Environment: The physical environment includes climate, landscape, soils, hydrology, local and regional topographic relief, and the geological processes that raise mountain ranges and create sedimentary environments. These processes evolved over a period of four billion years as a part of our planetary system and initially evolved with little or no organic influences. These variables play a key role in regulating the biological environment.

Geoarchaeology is an inherently interdisciplinary endeavor. Its evolution as a scientific discipline is lengthy, complex and interesting (for more thorough discussions of the discipline see Davidson and Shackley 1976; Herz and Garrison 1998; Goldberg et al. 2001; Holliday 1992; Lasca and Donahue 1990; Mandel 2000; Rapp and Gifford 1985; Rapp and Hill 1998; Stein 2001a; Stein and Farrand 1985; Waters 1992). Geoarchaeological contributions to SCAPE research by geoscientists fall within the broad definition of geoarchaeology provided by Goldberg et al. (2001: vii) as follows, "(geoarchaeology includes) any subject that bridges the interface between the earth sciences and archaeology, with the earth sciences including a wide array of subjects, such as sedimentology, stratigraphy, geomorphology, pedology, geochemistry, geophysics, and geochronology." As SCAPE research progressed, the terms geoarchaeology and geoarchaeologist were defined even more inclusively. In short, a SCAPE geoarchaeologist is a geoscientist willing to contribute to investigations aimed at improving the understanding of human-environment interaction in the past at temporal and spatial scales relevant to humans.

Role of the geoarchaeologist in the SCAPE project-The role of the geoarchaeologist in an interdisciplinary investigation such as SCAPE is to provide information about the physical conditions of the environment encountered by past human groups (e.g., Gladfelter 1977; Goldberg et al. 2001; Hassan 1979; Holliday 2004; Rapp and Hill 1998; Waters 1992). Typically, geoarchaeologists focus on four research questions (Gladfelter 1977). First, geoarchaeologists work to identify geologic processes responsible for archaeological site formation and burial (Stein 2001b). Second, geoarchaeologists identify the impact of natural processes that disturb or alter the primary position of archaeological material after abandonment by humans (Johnson 1990, 1989; Johnson and Hansen 1974; Leigh 2001; Stein 1983; Wood and Johnson 1978) or that affect the range of artifacts types (bone, textiles, pottery, stone, plant remains) likely to be preserved in an archaeological site (Wood and Johnson 1978). Third, geoarchaeologists assist in establishing stratigraphic relationships and chronostratigraphic control within sites (Stein 2000; Mandel and Bettis 2001) and across larger areas of interest (Ferring 2001; Frederick 2001; Holliday 2004, 1997, 1995).

This endeavor involves considerable investigation off-site (Holliday 2004), a circumstance that sometimes causes consternation among archaeologists who anticipated their geoscience colleagues would spend the bulk of their research time working at the archaeological sites under investigation. Fourth, and perhaps most importantly, geoarchaeologists are called upon to reconstruct the physical conditions of the environment, and the impact humans had on their environment, when humans were present at the archaeological sites under investigation (Bettis and Hajic 1995; Gladfelter 1977; Goldberg et al. 2001; Holliday 2004, 1997, 1992; Michlovic and Running In Press; Stein 2001a, 1995; Stein and Farrand 2001). More recently, geoarchaeologists are becoming increasingly involved in developing site predictive and settlement pattern models (Wells 2001) and sourcing of lithic resources (Herz 2001).

As SCAPE research, and the complexity of research questions under investigation evolved, a need to expand the range of available geoscience expertise was recognized. The cohort of participating geoscientists expanded accordingly (Running et al. this volume). It was recognized early that the geomorphic complexity characteristic of SCAPE study areas is in part inherited from glacial processes. Consequently, Quaternary geologists familiar with the Quaternary history of the study areas were consulted and meetings with such experts in the field were organized. Because aeolian landforms are an important component of the landscape of several of the SCAPE study areas, a geoscientist with expertise in aeolian sedimentology was added to the team. In addition, arrangements were made to meet and work with geoscientists with expertise in post-glacial aeolian records in the Canadian prairies and geochemical techniques to determine the age of aeolian sand dunes in the field. In an effort to expand the range of paleoenvironmental proxy-indicators we could apply to SCAPE research, geoscientists with expertise in the analysis of stable carbon isotopes and molluscan fossil assemblages were added. To assist in locating buried cultural remains over broad areas, an expert in the use of ground penetrating radar to investigate the near subsurface was engaged.

Temporal and Spatial Scales-Geoarchaeologists are called upon to paint a holistic picture of the physical conditions extant during the time period of interest including soils, landforms, plant and animal communities, and weather and climate. Geoarchaeologists must provide reconstructions of past environmental conditions at temporal and spatial scales relevant to humans (Gladfelter 1977; Stein 1993; Wandsnider 1998). Temporally, this means time scales that span human generations (about 50 years) and human memory-usually about five generations, although it is recognized that in some cases, oral traditions span centuries or even millenia. Spatially, this refers firstly to the site-specific scale, where humans conducted some activity or activities, and secondly the local-scale (an area defined by a concentration of archaeological sites or a suite of related landforms such as a glacial lake basin, a dune field, or a terminal moraine complex) where people conducted their daily lives (see also Wandsnider 1998). Geoarchaeologists are also called upon to link site specific and local scale paleoenvironmental reconstructions to reconstructions at the millennial and regional scales.

These finer scales of temporal and spatial resolution are sometimes difficult for geoarchaeologists to provide. In part this is because exogenic factors that control processes of landform and soil genesis such as climate and vegetation often vary, or are portrayed to vary, at the millennial and regional scales. Many proxy data sets used to infer paleoenvironmental conditions, particularly those from terrestrial settings such as soils and landforms, often do not routinely yield a record of environmental information at the decadal or even century scale. Other proxy data sets, particularly those recovered from subaqeuous settings (e.g., pollen records) tend to yield paleoenvironmental records at the regional-scale. In part, scale of inquiry problems reflects the research interests of many geoscientists who are interested in reconstructing paleoenvironments at these broader scales. Geoscientists filling the role of geoarchaeologists must recognize that providing paleoenvironmental reconstructions at these broader scales alone does not meet the needs of their archaeological colleagues.

Despite the difficulties, problems arising from differences in scales of inquiry between geoarchaeologists and archaeologists can be overcome and common ground can be found. For example, in soil geomorphology, an important component of many geoarchaeological investigations, the pedon, the smallest body of soil variability on the landscape, is similar in size to the archaeological excavation unit (Holliday 2004). This unit of analysis or a similar small unit of analysis can provide the kinds of fine scale information that was relevant to the thinking of the people that inhabited the site and the archaeologists that are attempting to understand it.

Geoarchaeological contributions to an interdisciplinary investigation such as SCAPE, require a team of geoscientists acting as geoarchaeologists. Each geoscientist uses the core concepts, theoretical background, and field and laboratory methods of their sub discipline.

Biological Environment-Archaeobotany and Plant Paleoecology With an understanding of the physical context in place, it is necessary to populate that landscape with living organisms. It is the plant and animal communities that enable human groups to survive and prosper over the generations. These communities include all living things in a region, from bacteria through plants, insects, fish, amphibians, reptiles, birds and mammals. These organisms are bound together in ecological relationships and into biological communities. Humans appeared late in this scheme but we are no less bound into these ecological relationships.

Biological activity influences the physical environment. Soils are a product of the biological community acting upon sediments. Climate and atmosphere are influenced by biological activity as well.

Human subsistence is dependent upon an adequate capture of energy and vital nutrients from the local environment. These local resources may be supplemented through trade and exchange with other groups having complementary surpluses. Technologies such as plant and animal domestication and other sophisticated harvesting practices can improve levels of energy capture. It is noted that, however technologically and socially sophisticated a group may become, their dependence upon biological resources remains as a fundamental requirement for their continued existence. Similarly, all living things, including humans, maintain themselves within the parameters of the physical environment.

With an ecological research orientation, the site in its landscape context can become the unit of analysis since the site, and its ancient occupants, can be considered in a temporally appropriate paleo-environmental context. This has led to an increased focus upon consideration of the subsistence requirements and seasonal use bases for site occupation. Land use selection criteria, the nature of the political economy implicit in land use, and how short and long term climatic change can affect human adaptation, are also basic considerations.

The study of past vegetation changes--the domain of plant paleoecology and archaeobotany--provides a crucial framework for understanding changing human-environmental interactions through time, and the impact of climate and other natural processes on biotic resources. Due to cross-disciplinary advances in paleoecology over the past two decades, a considerable number of vegetation proxies are available for application in archaeological research. These various proxies have well-documented strengths and weaknesses, and they are best employed in combination. Three key plant proxies (indicators), pollen, phytoliths, and plant macrofossils, used in SCAPE research are briefly summarized below.

Pollen-Palynology is a well-established method in paleoenvironmental research. Under anaerobic conditions, pollen grains become incorporated in sediment and preserve a record of changing abundance of plant species through time. Because plant communities may change in response to many different factors (both allogenic and autogenic), palynology is an important tool for deciphering complex, long-term, ecosystem processes. Systematic successional changes occurring at the regional level are usually assumed to represent the response of vegetation to shifts in mean climatic conditions (see Birks & Birks, 1980; MacDonald, 1990). Changes occurring at the stand level, on the other hand, may reflect the role of wildfires, soil, competition, and herbivory (among other factors) in structuring plant assemblage composition.

Although interpretation of pollen data is complicated by equifinality, differential production and preservation, and redeposition, a multiproxy (indicator) approach such as the one employed in this project enables production of robust paleoenvironmental models.

It could be argued that pollen, by itself, is poorly suited to the reconstruction of vegetation histories on the Northern Plains. This is because: first, pollen is rarely identifiable below the family level (e.g. Poaceae), and so changes in the distribution of mesic and xeric grass species through time are essentially invisible. This problem, to a large extent, is ameliorated through the analysis of grass phytoliths. Second, an important aspect of the ecological history of the Northern Plains relates to shifts in the location of the grassland-aspen forest boundary through time in response to climate and fire. However, this aspect of the regional environmental story is hampered by the poor preservation, and morphological indistinctiveness, of poplar (Populus) pollen. Importantly, however, this problem may be overcome through the analysis of soil and sediment for stable C isotope content.

This method was recently applied by Clark et al. (2001) to track shifts in the C4 (tall grass prairie)/ C3 (forest) boundary in Minnesota through the Holocene. Third, pollen analyses are also hampered, particularly in the western portion of the SCAPE study area, by poor preservation in the sedimentary record. Preservation in terrestrial settings such as archaeological sites is minimal (see Walker, 1992). By itself, therefore, palynology offers only moderate to minimal aid in developing the paleoecological context of archaeological sites on the Northern Plains.

Phytoliths-Although pollen and plant macrofossils are not typically preserved in ("dry") terrestrial soils, phytoliths are. Silicophytoliths (opaline phytoliths) are mostly composed of hydrated silicon dioxide ($SiO_2 * nH_2O$). Production occurs through the uptake of monosilicic acid from soil water and the subsequent precipitation (by evaporation and metabolism) of silica in the cell lumen and around the cell walls of the plant. Because the silica is deposited within the cellular matrix, it often assumes the size, shape, and configuration of the cell in which it forms (Geis, 1973: 113). Morphological variation in phytoliths may therefore be used as a means of associating particular forms with the plant taxa that produced them. Although phytoliths are also composed of calcium oxalate (see Jones and Bryant, 1992 for the Opuntia genus), siliceous forms are abundant in grasses.

Phytoliths hold enormous promise for the paleoecology of grasslands because this proxy provides information that is not available through conventional pollen analysis. For example, unlike grass pollen, which is only identifiable to the family level (Poaceae/ Gramineae), grass phytoliths are diagnostic to the subfamily (Twiss et al., 1969; Twiss, 1992). This is crucial, because different grass subfamilies have different preferences with regard to temperature and precipitation (e.g., Twiss, 1992). Consequently, changes in phytolith assemblages may potentially be used to reconstruct macroclimatic trends in grasslands (e.g., Cummings, 1996; Fredlund et al., 1998; Boyd, 2005). Secondly, although the pollen grains of many taxa are transported far from the source plant, phytoliths in soil are assumed by many to enter soil through in situ decay mechanisms (Dimbleby, 1978: 129; Piperno, 1988; Mulholland, 1989).

Phytolith assemblages may therefore be used to reconstruct local (rather than regional) vegetation composition, adding high spatial precision and "realism" to paleoecological models built using other proxies. Furthermore, as recent studies have shown, phytoliths may even be used to reconstruct long-term grassland fire histories, based on the inclusion of particulate carbon (Boyd, 2002), or changes in the refractive index for phytoliths exposed to high temperatures (Elbaum et al., 2003).

Plant Macrofossils-Plant macrofossils consist of any reproductive or vegetative component visible to the naked eye (e.g., seeds, wood, leaves, etc). This proxy offers insight into local-scale vegetation (the level of site or locale) and, because macrofossils may often be identified to the species level, offers more nuanced understanding of ancient plant assemblages (vs. pollen or phytoliths) where adequately preserved. In the SCAPE project, early Holocene macrofossil assemblages have been analyzed at several study areas, and these data have provided a useful local-scale context for Paleoindian sites (Boyd et al., 2003; Boyd, this volume). In particular, the macrofossil data has documented in detail the rapid colonization of wetlands by vegetation following the retreat of the Laurentide Ice Sheet. These productive wetlands were especially numerous on the fringes of proglacial lakes--which were nearly abiotic themselves--and it was to this 'productive fringe' that Paleoindian populations may have been attracted (Boyd et al., 2003; Boyd, this volume).

Examination of feature fill for macrobotanical food remains is a routine component of archaeobotanical research. In Southern Manitoba, considerable energy has been spent in the search for evidence of maize (Zea mays) consumption during the Plains Woodland period. Interest in this subject was first expressed in a series of earlier papers written by Nicholson (e.g., 1990, 1991, 1993, 1994, Nicholson et al. 2002). However, macrobotanical evidence of maize and other common cultigens has consistently failed to appear in features from Vickers Focus and other Plains Woodland sites in the province, and the archaeological record offers only moderate support for indigenous horticulture. Recently, however, this search was expanded to include phytolith analysis of features and carbonized food residue (on ceramic sherds) at the same sites. Unlike the macrobotanical studies, however, the phytolith research proved remarkably successful-diagnostic maize phytoliths have recently been identified in hearths and other features, and in the carbonized food residue (Boyd et al. 2006). The absence of corn macrofossils is interesting, and may reflect short (non-permanent) occupation of these sites, trading for maize instead of growing it on-site, or perhaps the relative rarity of features where these remains might be preserved in sufficient quantities (e.g., storage pits, burnt house structures, etc). In any case, this example underlines the importance of a multiproxy research strategy.

DISCUSSION

Plant paleoecology and archaeobotany, employing all three proxies identified above, have been a major focus of SCAPE activities. In order to address our research goals, the nature of SCAPE paleoecological research has, by necessity, differed in several ways from traditional paleoenvironmental approaches. Specifically, we have diverged from traditional 'paleoscience' by:

(1) focusing on unconventional paleoecological sites; (2) emphasizing both the response of human populations to environmental change as well as the potential impacts of small-scale human societies on their landscapes; and (3) focusing on sub-regional/ local environmental reconstruction. These points are briefly reviewed below.

Holocene vegetation changes on the Northern Plains are not well understood in contrast to many other regions of North America. This is because many lakes in the region tend to be ephemeral and saline and therefore not conducive to good pollen or organic macrofossil preservation, particularly during more arid periods. In many basins, furthermore, episodes of prolonged aridity prior to the late Holocene caused these lakes to dry up, and the compacted sediments formed during these events have rendered recovery of underlying sediments difficult (Barnosky et al., 1987; Beaudoin, 1993; Vance et al., 1993). Thus, the paleoecological record on the Northern Plains is therefore rather 'spotty', with some periods (e.g., the middle Holocene, ~8000-3000 14C yrs BP) and regions (e.g., SW Saskatchewan and SE Alberta) being particularly under-represented. In the past decade, important new insight into climate change on the Plains has been obtained through analysis of evaporite mineral content, isotopes, and diatoms (and other algae) in lake sediment (e.g., Last et al., 1998; Leavitt et al., 1999). However, these innovations provide no information on vegetation history. In the SCAPE project, despite the absence of deep, permanent, lakes in many of the study areas, a considerable amount of paleovegetation data was collected from 'unconventional' sites such as river cutbanks (e.g., Boyd et al., 2003; Boyd, accepted), aeolian soil sequences (Boyd, 2002, 2005), valley fill deposits (Boyd, in press), and dry kettle depressions. In many cases, these sites-while not providing continuous paleoecological records-yielded wonderfully detailed 'snapshots' of mid-and early-Holocene biota, which could be linked to the local archaeological record.

For example, in the glacial Lake Hind basin (SW Manitoba), erosion of a large parabolic dune by the Souris River exposed two zones of saturated organics (one dating to ~10,400-9300 14C yrs BP, the other dating to the mid-Holocene, ~6000-5000 14C yrs BP), which contained abundant plant macrofossils, pollen, and other organic remains. Aeolian soil sequences, furthermore, provided important insight into late Holocene prairie fire histories (Boyd, 2002) and local-scale vegetation succession (Boyd, 2005). In general, a multiproxy approach to paleoecological data collection was essential in order to maximize the amount of information possible from the variety of deposition settings encountered in the study areas.

Plains archaeologists have always been interested in the relationship between human populations and environment. However, little or no attention has been given to the modification of prairie environments by humans. Instead, most studies are concerned with tracking the responses of hunter-gatherer or village populations to environmental (especially climate) change (e.g., Walker, 1992)--reflecting, perhaps, the tragic Euroamerican experience during the 1930s "Dust Bowl". Although there can be little doubt that some changes in climate may have precipitated short-and long-term cultural responses, this is only part of the story. One goal of SCAPE research has been to identify the role, if any, of anthropogenic burning on vegetation in the Canadian Prairies (see Boyd, 2002; Boyd et al., in press; Oetelaar & Oetelaar, this volume). This objective is hampered by: (1) the difficulty of reconstructing fire histories in grassland environments; and (2) the difficulty in distinguishing natural from anthropogenic fire regimes.

This is an important area of study, however, because historical evidence consistently points to the use of fire by Plains hunter-gatherers to increase the predictability of bison herds (see Boyd, 2002). With regard to the first problem, new paleoenvironmental indicators of fire such as the presence of occluded carbon in phytoliths (Boyd, 2002), or changes in their refractive index (Elbaum et al., 2003), are promising avenues for further research. With regard to the second issue, Boyd (2002) has argued that archaeological evidence regarding changing Aboriginal land-use patterns should be considered in combination with standard paleoecological methods in order to address this problem in a convincing way. This is because prehistoric land-use was often focused on particular (central) areas, which changed through time in response to a variety of factors. Because the likelihood of anthropogenic burning may be expected to be higher in central (vs. peripheral) areas, data on prehistoric land-use are necessary to interpret changing fire regimes through time in a given area (Boyd, 2002). In general, multiple lines of evidence, and judicial consideration of both climatic and anthropogenic factors are necessary to understand the fire history of a given locale (see Boyd, 2002; Boyd et al., in press).

SCAPE paleoecological research has also departed from traditional approaches through its emphasis on local-scale (vs. regional) environmental reconstruction. The Canadian Prairies are a surprisingly heterogeneous region, containing a variety of landscapes, which, in turn, have unique biotic and geomorphic histories. In traditional paleoecological research, the desire to identify macroclimate change during the Holocene, and the impact of these changes on biota, requires a broad (regional) geographical scope. However, by itself, this scale conveys little insight for archaeology. Of particular interest in this project is the identification of areas of relatively high resource diversity and stability, which may have been preferentially occupied by hunter-gatherers throughout the Holocene (Boyd et al., in press; Kornfeld and Osborn 2003).

The goal of identifying these locales and tracing their histories requires the integration of modern vegetation surveys, presettlement landcover reconstructions, and subregional geomorphic and paleovegetation data. Using this approach for the Oak Lake Sandhills (MB), Boyd et al. (in press) suggest that this landscape differed from the surrounding drift prairie by higher ecosystem diversity and its greater resistance/ resilience to short-term climate change (e.g., late Holocene drought cycles). Plains Woodland site clusters in this region, therefore, may reflect multigenerational, preferential. selection of this landscape by hunter-gatherers seeking to reduce subsistence risk (Boyd et al, in press).

ARCHAEOLOGY

Many species, particularly mammals, have developed effective social configurations that enhance their collective survival within their particular environmental contexts. In some cases, many of their behaviors appear to be genetically based. With humans this is not the case except for a few innate responses such as the nursing impulse in infants. Humans have evolved a complex suite of flexible learned behaviors that we subsume under the term culture. The recent human cultural adaptation is enabled by a command of spoken language that uses arbitrary vocal symbols to communicate abstract ideas, anticipated events in the future, and remembered events in the past. While some degree of learning has been demonstrated for non-human species, innate instinctive behavior dominates the responses of other animals to their environment. Changing such behavior, in response to environmental changes, usually requires progressive responses over several generations. Humans, in contrast, can modify collective responses to environmental challenges or opportunities extremely rapidly.

In attempts to interpret the subsistence practices and socio-economic activities of past groups, archaeologists have developed numerous analytical techniques based upon the material culture remains recovered from archaeological sites.

The within-site context of these remains has become, over the past several decades, a major methodological focus with precise three-dimensional provenience of features and artifacts being the expected norm for recovery.

This within-site context includes taphonomic studies that deal with a range of vectors that disturb original patterns of deposition and that lead to the disintegration of some classes of material culture-particularly organic materials such as wood and bone (Lyman 1994). The larger questions of context have often been ignored or scavenged from the literature of other disciplines with sometimes less than perfect results. This is particularly the case when modern studies by botanists and ecologists have been uncritically accepted as usable models for an environment of hundreds or thousands of years ago that may have been very different from the present situation.

Among the earliest attempts at analysis of material culture, the definition and description of the "diagnostic artifact" has been preeminent. Archaeological Cultures are defined in this way and culture history and settlement pattern studies rely on this kind of analysis as a fundamental given. The Clovis point and the Blackduck pot immediately resonate with anyone interested in placing the people who made these artifacts in time and space. However, that is about as far as one can go with that level of analysis. In addition, many of the diagnostic artifacts are somewhat ambiguous, particularly if they have been rejuvenated or modified subsequent to their original fabrication. Walker (1992-141) notes that, "In comparison to other Early Middle Period point types, the Gowen Side-notched point is relatively nondescript." Retouched Gowen points, recovered from an undated context for example, vary so widely that they could be assigned to Gowen, Besant, Oxbow, Sandy Creek or other standardized types on a case-by-case basis (see also Walker 1992:45 & 76).

Functional analyses of material culture (Kooyman 2000; Rice 1987) can identify a variety of activities that were conducted at a given site. Similarly, use-wear (Hayden 1979; Keeley 1980) and residue analysis (Boyd et al. 2006b; Quigg et al. 2001) can expand our knowledge of the activities conducted at a given site.

Zooarchaeological analyses have been successfully used to infer season of occupation of sites, hunting and butchering practices and, in some cases, to suggest patterns of cultural distribution within sites (Lyman 1994; Playford 2001). Features, such as hearths and storage pits also provide focal points for specific activities through associated artifact clusters or preserved organic materials.

Social organization and kinship structures are notoriously difficult to infer for precontact groups. However, ethnohistoric and early ethnographic accounts can provide useful data for hypothesis development and occasionally for direct interpretation-particularly for the most recent time periods when a direct cultural affiliation may be plausible (e.g. MacNeish 1958:67). Another example would be the identification of a divination technique (Nicholson 1978) used at the precontact Aschkibokahn site on an island where the Duck and Drake rivers empty into Lake Winnipegosis. In this case, an ethnographic account (Honigmann 1956) provided a rationale for a non-functional pattern of bone breakage in beavers. Similarly, oral traditions that have been passed from generation to generation sometimes offer direct accounts of relevant events that are useful for particular cases.

SUMMARY

To varying degrees, the biological and cultural environments interact with the physical aspects of the environment and modify the physical processes. The biological and cultural aspects also interact with each other in a dynamic way, placing limits on each other, and causing changes in natural and human ecological outcomes.

Through a range of techniques drawn from the Biological and Earth Sciences, the Social Sciences and the Humanities, we have been able to establish broad contextual milieus for the interpretation of archaeological sites and the lifeways of the people that occupied them, across a time period spanning over 6,000 years. The contained knowledge and expertise of a single discipline could not have achieved these results.

10

Landscape Ecology

Landscape ecology is a sub-discipline of ecology and geography that address how spatial variation in the landscape affects ecological processes such as the distribution and flow of energy, materials and individuals in the environment (which, in turn, may influence the distribution of landscape "elements" themselves such as hedgerows). Landscape ecology typically deals with problems in an applied and holistic context.

TERMINOLOGY

The term landscape ecology was coined by Carl Troll, a German geographer in 1939 (Troll 1939). He developed this terminology and many early concepts of landscape ecology as part of his early work applying aerial photograph interpretation to studies of interactions between environment and vegetation.

EXPLANATION

Heterogeneity is the measure of how different parts of a landscape are from one another. Landscape ecology looks at how spatial structure affects organism abundance at the landscape level, as well as the behavior and functioning of the landscape as a whole. This includes the study of the pattern, or the internal order of a landscape, on process, or the continuous operation of functions of organisms (Turner 1989). Landscape ecology also includes geomorphology as applied to the design and architecture of landscapes (Allaby 1998). Geomorphology is the study of how geological formations are responsible for the structure of a landscape.

HISTORY

Evolution of Theory

One central landscape ecology theory originated from MacArthur & Wilson's The Theory of Island Biogeography. This work considered the assembly of flora and fauna on islands as the result of colonization from a mainland stock and stochastic extinction. The concepts of island biogeography were generalized from physical islands to abstract patches of habitat by Levins' metapopulation model. This generalization spurred the growth of landscape ecology by providing conservation biologists a new tool to assess how habitat fragmentation affects population viability. Recent growth of landscape ecology owes much to the development of geographic information systems (GIS) technology and the availability of large-extent habitat data (e.g. remotely sensed satellite images or aerial photography).

Development as a Discipline

Landscape ecology developed in Europe from historical planning on human-dominated landscapes. In North America, concepts from general ecology theory were integrated. While general ecology theory and its sub-disciplines focused on the study of more homogenous, discrete community units organized in a hierarchical structure (typically as populations, species, and communities), landscape ecology built upon heterogeneity in space and time, and frequently included human-caused landscape changes in theory and application of concepts (Sanderson and Harris 2000).

By 1980, landscape ecology was a discrete, established discipline, marked by the organization of the International Association for Landscape Ecology (IALE) in 1982 and landmark book publications defining the scope and goals of the discipline, including Naveh and Lieberman (1984) and Forman and Godron (1986) (Ryszkowski 2002). Forman (1995) wrote that although study of "the ecology of spatial configuration at the human scale" was barely a decade old, there was strong potential for theory development and application of the conceptual framework. Today, theory and application of landscape ecology continues to develop through a need for innovative applications in a changing landscape and environment. Landscape ecology today relies more on advanced technologies such as remote sensing, GIS, and simulation models, with associated development of powerful quantitative methcds to examine the interactions of patterns and processes (Turner et al. 2001). An example would be determining the amount of carbon present in the soil based on landform over a landscape, derived from GIS maps, vegetation types, and rainfall data for a region.

Relationship to Ecological Theory

Although landscape ecology theory may be slightly outside of the "classical and preferred domain of scientific disciplines" because of the large, heterogeneous areas of study (Sanderson and Harris 2000), general ecology theory is central to landscape ecology theory in many aspects. Landscape ecology is comprised of four main principles, which include: 1. the development and dynamics of spatial heterogeneity, 2. interactions and exchanges across heterogeneous landscapes, 3. influences of spatial heterogeneity on biotic and abiotic processes, and 4. the management of spatial heterogeneity. The main difference from traditional ecological studies, which frequently assume that systems are spatially homogenous, is the consideration of spatial patterns (Turner and Gardner 1991).

Important Terms in Landscape Ecology

Landscape ecology not only embraced a new vocabulary of terms but also incorporated general ecology theory terms in new ways. Many of the terms used in landscape ecology are as interconnected and interrelated as the discipline itself. Landscape can be defined as an area containing two or more ecosystems in close proximity (Sanderson and Harris 2000).

Scale and Heterogeneity (incorporating composition, structure, and function)

A main concept in landscape ecology is scale. Scale represents the real world as translated onto a map, in the relationship between distance on a map image and the corresponding distance on earth (Malczewski 1999). Scale is also the spatial or temporal measure of an object or a process (Turner and Gardner 1991), or level or degree of spatial resolution (Forman 1995). Components of scale include composition, structure, and function, which are all important ecological concepts. Applied to landscape ecology, composition refers to the number of patch types (see below) represented on a landscape, and their relative abundance. For example, the amount of forest or wetland, the length of forest edge, or the density of roads can be aspects of landscape composition.

Structure is determined by the composition, the configuration, and the proportion of different patches across the landscape, while function refers to how each element in the landscape interacts based on its life cycle events (Turner and Gardner 1991). Pattern is the term for the contents and internal order of a heterogeneous area of land (Forman and Godron 1986).

A landscape with structure and pattern implies that it has spatial heterogeneity, or the uneven, non-random distribution of objects across the landscape (Forman 1995). Heterogeneity is a key element of landscape ecology that separates this discipline from other branches of ecology.

Patch and Mosaic

Patch, a term fundamental to landscape ecology, is defined as a relatively homogeneous area that differs from its surroundings (Forman 1995). Patches are the basic unit of the landscape that change and fluctuate, a process called patch dynamics. Patches have a definite shape and spatial configuration, and can be described compositionally by internal variables such as number of trees, number of tree species, height of trees, or other similar measurements (Forman 1995).

Matrix is the "background ecological system" of a landscape with a high degree of connectivity. Connectivity is the measure of how connected or spatially continuous a corridor, network, or matrix is (Forman 1995). For example, a forested landscape (the matrix) with fewer gaps in forest cover (open patches) will have higher connectivity. Corridors have important functions as strips of a particular type of landscape differing from adjacent land on both sides (Forman 1995). A network is an interconnected system of corridors while mosaic describes the pattern of patches, corridors and matrix that form a landscape in its entirety (Forman 1995).

Boundary and Edge

Landscape patches have a boundary between them which can be defined or fuzzy (Sanderson and Harris 2000). The zone composed of the edges of adjacent ecosystems is the boundary (Forman 1995). Edge means the portion of an ecosystem near its perimeter, where influences of the adjacent patches can cause an environmental difference between the interior of the patch and its edge. This edge effect includes a distinctive species composition or abundance in the outer part of the landscape patch (Forman 1995). For example, when a landscape is a mosaic of perceptibly different types, such as a forest adjacent to a grassland, the edge is the location where the two types adjoin. In a continuous landscape, such as a forest giving way to open woodland, the exact edge location is fuzzy and is sometimes determined by a local gradient exceeding a threshold, such as the point where the tree cover falls below thirty-five percent (Turner and Gardner 1991).

Ecotones, Ecoclines, and Ecotopes

A type of boundary is the ecotone, or the transitional zone between two communities (Allaby 1998). Ecotones can arise naturally, such as a lakeshore, or can be human-created, such as a cleared agricultural field from a forest (Allaby 1998). The ecotonal community retains characteristics of each bordering community and often contains species not found in the adjacent communities. Classic examples of ecotones include fencerows; forest to marshlands transitions; forest to grassland transitions; or land-water interfaces such as riparian zones in forests. Characteristics of ecotones include vegetational sharpness, physiognomic change, occurrence of a spatial community mosaic, many exotic species, ecotonal species, spatial mass effect, and species richness higher or lower than either side of the ecotone (Walker et al. 2003).

An ecocline is another type of landscape boundary, but it is a gradual and continuous change in environmental conditions of an ecosystem or community. Ecoclines help explain the distribution and diversity of organisms within a landscape because certain organisms survive better under certain conditions, which change along the ecocline. They contain heterogeneous communities which are considered more environmentally stable than those of ecotones (Attrill and Rundle 2002).

An ecotope is a spatial term representing the smallest ecologically-distinct unit in mapping and classification of landscapes (Forman 1995). Relatively homogeneous, they are spatially-explicit landscape units used to stratify landscapes into ecologically distinct features for measurement and mapping of landscape structure, function, and change over time, and to examine the effects of disturbance and fragmentation.

Disturbance and Fragmentation

Disturbance is an event that significantly alters the pattern of variation in the structure or function of a system, while fragmentation is the breaking up of a habitat, ecosystem, or land-use type into smaller parcels (Forman 1995). Disturbance is generally considered a natural process. Fragmentation causes land transformation, an important current process in landscapes as more and more development occurs.

LANDSCAPE ECOLOGY THEORY

Elements of Landscape Ecology Theory

Landscape ecology, as a theory, stresses the role of human impacts on landscape structures and functions and proposes ways for restoring degraded landscapes (Naveh and Lieberman 1984). Landscape ecology explicitly includes humans as entities that cause functional changes on the landscape (Sanderson and Harris 2000). Landscape ecology theory includes the landscape stability principle, which emphasizes the importance of landscape structural heterogeneity in developing resistance to disturbances, recovery from disturbances, and promoting total system stability (Forman and Godron 1986).

This principle is a major contribution to general ecological theories which highlight the importance of relationships among the various components of the landscape. Integrity of landscape components helps maintain resistance to external threats, including development and land transformation by human activity (Turner et al. 2001). Analysis of land use changes has included a strongly geographical approach within landscape ecology. This has led to acceptance of the idea of multifunctional properties of landscapes (Ryszkowski 2002). There are still calls for a more unified theory of landscape ecology due to differences in professional opinion among landscape ecologists, and the interdisciplinary approach to the discipline (Bastian 2001).

An important related theory is hierarchy theory, which refers to how systems of discrete functional elements operate when linked at two or more scales. For example, a forested landscape might be hierarchically composed of drainage basins, which in turn are composed of local ecosystems or stands, which are in turn composed of individual trees and tree gaps (Forman 1995).

Recent theoretical developments in landscape ecology have emphasized the relationship between pattern and process, as well as the effect that changes in spatial scale has on the potential to extrapolate information across scales (Turner and Gardner 1991). Several studies suggest that the landscape has critical thresholds at which ecological processes will show dramatic changes, such as the complete transformation of a landscape by an invasive species with a small change in average temperatures per year which favors the invasive habitat requirements (Turner and Gardner 1991).

LANDSCAPE ECOLOGY APPLICATION

Research Directions

Developments in landscape ecology illustrate the important relationships between spatial patterns and ecological processes, and incorporate quantitative methods that link spatial patterns and ecological processes at broad spatial and temporal scales. This linkage of time, space, and environmental change can assist land managers in applying land management plans to solve environmental problems (Turner et al. 2001).

The increased attention in recent years on spatial dynamics has highlighted the need for new quantitative methods that can analyze patterns, determine the importance of spatially explicit processes on the landscape, and develop reliable landscape models

(Turner and Gardner 1991). Multivariate analysis techniques, a type of statistics incorporating many variables, are frequently used to examine landscape level vegetation patterns. A number of studies in riparian systems and wetlands use a variety of statistical techniques, such as cluster analysis, canonical correspondence analysis (CCA), or detrended correspondence analysis (DCA), for classifying vegetation. Gradient analysis is another way to determine the vegetation structure across a landscape, or to help delineate critical wetland habitat for conservation or mitigation purposes (Lyon and Sagers 1998, Choesin and Boerner 2002).

Climate change is another major component in structuring current research in landscape ecology. Ecotones, as a basic unit in landscape studies, may have significance for management under climate change scenarios, since change effects are likely to be seen at ecotones first because of the unstable nature of a fringe habitat (Walker et al. 2003).

Research in northern regions has examined landscape ecological processes, such as the accumulation of snow during winter, snow melting, freeze-thaw action, percolation, soil moisture variation, and temperature regimes through long-term measurements in Norway (Loffler and Finch 2005). The study analyzes gradients across space and time between ecosystems of the central high mountains to determine relationships between distribution patterns of animals in their environment. Looking at where animals live, and how vegetation shifts over time, may provide insight into changes in snow and ice over long periods of time across the landscape as a whole.

Other landscape-scale studies maintain that human impact is likely the main determinant of landscape pattern over much of the globe (Wilson and King 1995). Landscapes may become substitutes for biodiversity measures because plant and animal composition differs between samples taken from sites within different landscape categories. Taxa, or different species, can "leak" from one habitat into another, which has implications for landscape ecology. As human land use practices expand, and continue to increase the proportion of edges in landscapes, the effects of this leakage across edges on assemblage integrity may become more significant and important in conservation because taxa may be conserved across landscape levels, if not at local levels (Dangerfield et al. 2003).

RELATIONSHIP TO OTHER DISCIPLINES

Landscape ecology has important links to application-oriented disciplines such as agriculture and forestry. In agriculture, landscape ecology has introduced new options for the control and management of environmental threats brought about by the intensification of agricultural practices. Agriculture has always been a strong human impact on ecosystems (Ryszkowski 2002). In forestry, changes in consumer needs have changed conservation and use of forested landscapes from structuring stands for fuelwood and timber to ordering stands across landscapes to enhance aesthetics, habitats and biological diversity. Landscape forestry provides methods, concepts, and analytic procedures for shifting management from traditional to landscape forestry (Boyce 1995). Landscape ecology has been cited as a major contributor to the development of fisheries biology as a distinct biological science discipline (Magnuson 1991), and is frequently incorporated in study design for wetland delineation in hydrology (Attrill and Rundle 2002).

11

Prehistoric, Historic, and Present Settlement Patterns Related to Ecological Hierarchy

The distribution of human occupation across a landscape provides information about how people use the landscape, about patterns of economic development, and about social interactions of human groups. When the distributions are examined over several thousand years, we gain an evolutionary understanding, not only of the people and their cultural patterns, but also of physical landscape development.

The focus of this assessment was to examine and compare settlement patterns of prehistoric, historic, and present time periods, based on known cultural sites in the Eastern Upper Peninsula of Michigan, U.S.A., and to generate hypotheses about the interaction of settlement pattern and landscape change at multiple scales. Patterns of settlement among the three time periods were compared at three geographic scales: by subregional ecosystems, by landscape ecosystems and by terrain characteristics.

The Michigan Bureau of History database of archaeological sites was searched for prehistoric habitation sites of Middle or Late Woodland period (ca. 3000-300 years before present). Historic occupations were drawn from pre-European settlement landscape data based on General Land Office survey notes of the 1850s.

We extracted "urban" categories from land cover classified from Landsat Thematic Mapper imagery to measure present occupations. Spatial patterns and dynamics of settlement areas in each time period were examined using the ARC/INFO geographic information system (GIS). Results showed a tendency for settlement in all time periods on the bedrock and lowland landscape groups near Great Lakes shorelines, generally occupying slopes less than two percent. The distribution of present occupations, in terms of both slope aspect and geographic subregion (multi-scalar), was similar to the distribution of prehistoric occupations. Both prehistoric and present sites were primarily south facing and were frequently found along Green Bay and Lake Michigan shorelines.

Likewise, the interaction of landscape structure and cultural patterns provides information to the fields of archaeology, anthropology, ethnography, and cultural geography. As societies modify their environment in characteristic ways, they reflect aspects of their technology and social interaction (Kelso 1993; Nassauer & Westmacott 1987). Benefits of applying theories from geology and landscape ecology toward understanding the paleo and historic landscape record have been demonstrated by Hammett (1992), McGlade (1995), Stafford (1995), and Rossignol & Wandsnider (1992).

For instance, McGlade (1995), claimed the theoretical anthropology debate is "missing an adequate appreciation and understanding of the way in which social practice inserts itself within the natural world and shares with it a reciprocal dynamic." At the same time, studies are needed to integrate human cultural and landscape ecological theory (Golley 1995).

Unfortunately, research that probes the interaction of cultural and biophysical dynamics has been more of an aspiration than an accomplishment among landscape ecologists (Nassauer 1995).

Knowledge of how dynamics operating at varying hierarchical levels elucidate settlement preferences is still sparse. The emphasis of this paper was to generate hypotheses about these relationships and to establish a multi-scale framework of settlement/ landscape patterns on which finer scale examinations can take place. The specific objectives of this paper are to: 1) examine human occupation distributions of prehistoric (Woodland), historic (1850s), and present (1990s) settlement patterns; and 2) compare settlement distribution among the three time periods by subregional ecosystems, landscape ecosystems and terrain.

STUDY AREA

The Eastern Upper Peninsula of Michigan lies between three Great Lakes; Huron, Michigan, and Superior, resulting in lake-moderated temperatures and high precipitation rates (Map 1). The region is characterized by a variety of land types including; hardwood forest on glacial moraines and limestone outcrops, jack pine barrens on outwash and beach ridges, coastal marshes, interior bogs, fens and cedar swamps (Peterson 1986). The area forms the northeastern portion of the Northern Great Lakes Ecological Section, 212H (Bailey et al. 1994).

DOCUMENTATION OF EARLY SETTLEMENTS

The last glacial front in the Upper Peninsula (the Marquette Advance) dissipated around 9,500 years before present (B.P.) (Peterson 1986). The earliest evidence of human occupation recorded in the area is the Gorto site, a possible Paleo-Indian (ca. 10,000-8,500 B.P.) site near Ishpeming. Early and Middle Archaic (ca. 8000-4000 B.P.) sites are not known for eastern northern Michigan, however, a Late Archaic (ca. 4000-3000 B.P.) site is located in the St. Ignace area. Early Woodland occupations are lacking in the area but Middle and Late Woodland sites are well documented (Anderton 1993, Benchley et al. 1988). "Woodland" refers to a cultural period from about 3000 B.P. to 300 years B.P., or the time of first European contact and historic documentation. In the Great Lakes region, the Woodland period is associated with the advent of pottery containers and agricultural practices. Tanner's atlas of Great Lakes Indian history (1987), which is based on documentary sources, showed Indian villages at L'Anse, Sault Ste. Marie, and St. Ignace as early as 1720. By 1768 new occupations had appeared at Grand Island, Nahma, Manistique, and Whitefish Point.

Studies of Woodland and Ojibwa people in Upper Michigan began with writings and journals of early European missionaries and travelers such as Allouez, Marquette, and Henry W. Schoolcraft (1970). In the early to mid 20th century, several ethnographic histories were written. Johann Kohl (1985) spent the summer of 1855 with Lake Superior Ojibwa. Frances Densmore (1974) studied Ojibwa people of northern Minnesota from 1907 to 1925, and

Kineitz (1947, 1991) lived in an Ojibwa village in the Western Upper Peninsula of Michigan. Finally, with construction of the Mackinac Bridge, researchers from the University of Michigan in the 1950s and 60s conducted the first systematic archaeological surveys in the Upper Peninsula of Michigan along the south shore of Lake Superior (Bigony 1968). Other systematic surveys have since been completed by Michigan State University and the Hiawatha National Forest on national forest lands (Dunham & Branstner 1995, Lovis 1979), and by Michigan Technological University and the Department of Transportation along the north shore of Lake Michigan (Martin & Martin 1979). As a result of these previous archaeological and ethnohistoric studies which have provided a record of human settlements, we were able to examine land use patterns spatially throughout a period of several thousand years.

METHODS

Prehistoric Sites

The use of paleobotany has been well-documented for assessing ecological patterns and processes where no written historical analogs exist. For example, pollen core samples of Great Lakes bogs were analyzed to record Holocene vegetation patterns (Graumlich and Davis 1993). Likewise, archaeological data can be used to capture settlement patterns across the region prior to written records. Documented Woodland settlements comprised the prehistoric settlement coverage. The Michigan Bureau of History database of archaeological sites was searched for sites meeting selection criteria within the study area. These criteria favored sites with explicit locations, evidence of Middle or Late Woodland period occupation (3000-300 years before present), and habitations versus other special use sites such as cemeteries, quarries, etc. Onehundred twelve sites met the criteria.

Because the database provided only point locations, an area of five hectares was assigned to the prehistoric sites for spatial comparisons. Five hectares were chosen because of the size of the historic cultural sites and because it is a scale at which the effective local environment of sites can be quantified. As a result of overlapping settlement point locations, closely adjacent occupations formed one settlement polygon. Eighty polygons resulted from this data manipulation.

Historic Cultural Features

In the United States, the General Land Office (GLO) survey reports are a common historic record used in scientific studies. The Michigan GLO surveyors took detailed notes not only on the location, species, and diameter of each tree used to mark section lines and corners, but also on natural and cultural features such as lakes, wetlands, fields, burned areas, blowdowns, trails, and settlements. The records have been used to reconstruct historic vegetation patterns (Veatch 1959) and for quantitative analysis of historic vegetation (Bourdo 1956; Curtis 1959; Canham 1978; Canham & Loucks 1984; Lorimer 1977, 1980; and Whitney 1986), examination of "old forest" (Frelich 1995), stand structure and composition (Palik and Pregitzer 1992; Price 1994), and disturbance frequencies (Baker 1991; Frelich & Lorimer 1991; and Lorimer 1984). Besides vegetation studies, the records can be used for cultural history. Historic cultural sites were based on these notations from the mid-1800s. The GLO maps and notes were compiled by Michigan Natural Features Inventory (MNFI) on 1:24,000 USGS topographic maps.

Forty-nine historic cultural sites occurred within the study area. Most sites were less than one hectare in size, several were between five and 40 hectares, and one, near the present location of Sault Ste. Marie, was over 1400 hectares.

Present Settlements (urban areas)

Present urban areas were obtained from Landsat Thematic Mapper (TM) imagery for the Upper Peninsula of Michigan from Eosat, Inc. (Maclean 1994). Of the several land cover classes separated in the TM imagery classification, one category, identified as "urban", was used to represent present settlements by drawing all "urban" coded polygons from the data set. The classified TM land cover data set was available in county-wide coverages. Due to the coverage size, the urban polygons were extracted from each county file prior to joining the six county coverages. A total of 725 urban sites resulted, the smallest was 1.5 hectares and the largest was 845 hectares.

Subregional Ecosystems

The subregional framework used to assess settlement distribution was a product of regionalization, a means of organizing the environment into understandable and measurable units for purposes of management and analysis (Allen & Starr 1982; Klijn & Udo de Haes 1994; Urban et al. 1987). Rowe (1991) described regionalization as partitioning landscapes into smaller areas that are increasingly more homogeneous based on dominant features which differentiate them from surrounding landscapes. Boundaries occur as perceived changes in the ecological relationships of vegetation, landform, drainage soil, and climate inferred from these patterns, leading to classification of areas into discrete units with boundaries that may actually represent very broad ecotones (Rowe 1996). Because management issues occur at multiple spatial scales, and because ecosystems are linked geographically, ecosystem mapping and analysis must be accomplished at different levels, from local (site) to broad groups of geographically related sites (landscapes and ecoregions) (Bailey 1996).

In the Upper Great Lakes region, common systems of hierarchical ecological units are being used for planning and analysis. These hypothetical structures facilitate systematic stratification and testing of processes and responses, and become the organizational units around which data and knowledge are accumulated (Uhlig and Jordan 1996). At the broadest scale, Bailey (1996) mapped continental domains, within which finer scale ecoregions (provinces and sections) were nested (Bailey et al. 1994). In the Lake States, Albert et al. (1986, 1995) mapped subregional ecosystems within Bailey's sections, based on climatic and broad physiographic factors. The subregional ecosystems are termed "subsections," and were used for the first level of settlement pattern analysis in this study.

Landtype Ecosystems

Albert's subsections provided the framework for subdivision into landscape ecosystem units called "landtype associations," characterized by topography, microclimate, soil, and vegetation (Albert et al. 1986). Recognition of pattern and distinctive-ness at one scale simply begins the delineation process, uncovering contradictions whose resolution leads to examination at new scales, where new elements and structures can be seen (Crumley 1994). Delineation of landtype associations was based on: terrain (major slope and elevation changes or escarpments, variability of slope) from 1:250,000 USGS topography maps; geology (bedrock type, surficial geology, major glacial features, paleoshorelines) from Anderton (1993), Martin (1957) and Farrand (1982) at 1:500,000; soil (communities of soils, drainage, texture, depth to bedrock, source and mode of deposition) from county-level soil association maps; localized climate (lake effect wind-blown sand and wave action, major cold air drainages, etc.) from Santer (1977); large lakes or open water systems; watersheds (drainage pattern, water level and fluctuation variability); and vegetation (LANDSAT 1984 image, historic vegetation from General Land Office surveyors notes, aerial photography at various scales); existing subsection maps; plot data; and personal experience (Albert 1995; Jordan et al. in prep.). Because the LTA units represent hypothetical boundaries between similar and dissimilar areas, map error and accuracy were not measurable. Rather, landscape pattern and composition, landscape change, and landscape process studies are used to quantify and describe the working map units for further analyses (Silbernagel 1996).

The landscape units were grouped into eight physiographic categories: morainal, outwash, lowland sand lake plain, bedrock controlled, clay lake plain, ground moraine, transitional/complex, and inland lake (Gary et al. 1974), comparable to Michigan's Land Types mapped by Michigan Department of Natural Resources (Santer 1977).

Terrain Data

The third perspective from which we assess habitation was by major terrain features. Digital elevation model (DEM) data from the US Geological Survey were examined in relation to settlement sites of the three time periods to generate hypotheses regarding differences in settlement orientation and location on major landforms. Resolution of this coverage was three by three arc seconds (latitude/ longitude) or about 60 by 90 square meters at latitudes of ca. 46 degrees north, with 1201 elevations per one by one degree profile.

Analytical Methods

The described coverages were spatially analyzed using workstation ARC/INFO v. 6.1.2 (Environmental Systems Research Institute, Inc. 2nd Ed., March 1992). The various coverages were overlaid and output reports from the INFO module were generated. The output reports were then imported to SAS (SAS Institute, Inc.) for statistical analyses. The original digital elevation data were in lattice form. To develop slope, aspect, and elevation measures, the lattice was converted to polygons using an ARC command, "latticepoly."

RESULTS

Distribution of Settlement Sites by Subsection

We examined settlement distribution of the three periods by subsection to assess broad geographic affinities. In all three time periods, subsection 212Hj, comparable to Albert's (1995) "St. Ignace" map unit along the Lake Michigan shoreline, was frequently used. Both prehistoric and present settlements also utilized subsection 212He, Albert's (1995) "Escanaba/Door Peninsula" along the Green Bay shoreline extensively, but this was not the case in historic times. Subsection 212Hi, "Grand Marais Sandy End Moraine and Outwash" (Albert 1995), bordering Lake Superior, was occupied frequently by prehistoric and historic villagers, but seldom by present inhabitants. Present developments were more common within subsections 212Hh, "Seney Sand Lake Plain" (Albert 1995) and 212Hk "Rudyard" clay plains area (Albert 1995).

Distribution of Settlement Sites by Landtype Association

Landtype associations most frequently used by prehistoric people were the bedrock and lowland groups. The historic data, with fewer total sites, occurred on the same LTA groups but with a slightly greater percentage on the transitional and clay plains groups, and with no documented occurrences on the outwash group. The present settlements occurred most frequently on the lowland group, secondly on the transitional group, and then on bedrock on clay plains groups.

Examination of settlement area by LTA group found that the historic data were skewed to the clay plains group, as a result of one exceptionally large (1408 ha) site in the historic data set. The site is located at the current location of Sault Ste. Marie. It was also evident that prehistoric sites have a relatively consistent mean area because they were all buffered to the same size. In contrast, the present site distribution by area was fairly representative of its frequency distribution.

Distribution of Settlement Sites by Terrain

We overlaid DEM data to describe site distribution by terrain variables such as aspect. Prehistoric data indicated a preference for south facing sites (160-200 degrees), whereas the historic villages were situated least frequently to the south and most frequently to the east (22-112 degrees). Present towns and sites, like prehistoric, favored a south facing aspect.

In Upper Michigan, 192 m is the current elevation of the Glacial Lake Nipissing shoreline of about 4500 years before present, which occurs as a prominent bench around much of Lake Michigan and Superior. Woodland sites were located just below the Nipissing bench (186 m), close to the shoreline, but by historic time villages were higher at around 200 m. The most heavily developed elevation in present times was around 192 m, then decreased as elevation increased. There were no recorded secondary clusters of present sites at higher elevations as there were with the prehistoric and historic sites. The slope graph (Fig. 6) portrayed all three datasets with like distributions, having most frequent occurrences on the lowest slope categories. In the prehistoric coverage, 95.2% of the settlements occurred on slopes less than 2.15%. The historic coverage contained 83.2% on slopes less than 2.15% and present settlements, almost 95%.

ASSUMPTIONS AND LIMITATIONS

Each settlement data layer had unique limitations which should be discussed. For example, in the prehistoric site coverage, selected sites were buffered to the chosen size of five hectares based on relevant information described previously. This size determination may not accurately represent variations in the size of the effective local environment of prehistoric sites. In studying Late Woodland sites, Holman (1978) discovered that effective local environment actually varied with the "shape" of the land. Hammett (1992) reviewed historic documents depicting Southeast landscapes and found them to be "various types of resource patches concentrated in concentric rings with areas of lesser exploitation spreading away from occupation centers," but does not provide area estimation for the primary exploitation area. The challenge lies in deciding what portion of the concentric resource areas constitute the "primary" exploitation area. Furthermore, the size of a "site" will affect where it falls within a terrain feature or landscape unit.

A small site would be specific to one aspect but a larger site might cover several. As with most ecological and archaeological studies, sampling completeness of the datasets must be considered since they affect the number and distribution of sites. The prehistoric site data was based on archaeological surveys, which were more intensive in certain areas within the Upper Peninsula than others. Lands along the lakeshores and within the National Forest generally received more archaeological investigations than other inland areas. The historic cultural site coverage, on the other hand, was based on the equivalent of a systematic aligned sampling strategy, a satisfactory method for recording a sample of standing cultural features.

The surveyors walked and recorded sites along the township and section survey lines (one mile intervals). But, the transcribed digital data from MNFI, contained no attribute information about these cultural sites, thus, original GLO surveyors maps and notes were examined for more information about the individual sites (Trygg 1964, 1969). The present settlement coverage was derived from a third type of methodology which, unlike the first two, essentially counted sites in the total study area rather than a sample. However, it appeared that areas which fell into the present settlement (urban) category based on their spectral signature were any areas with flat, highly absorptive surfaces. Consequently, this included pavement and bare rock as well as buildings and towns.

The terrain analysis was based on digital elevation data with relatively low resolution (60 x 90 m) which limited its value as an analysis tool. Hammer et al., (1995) found within a 1.6 by 2.4 km study area in Missouri, even 30 m resolution DEM only correctly classified slope classes 21-30% of the time. However, the spatial extent in this study was much larger (approximately 100 km by

250 km) therefore, a lower resolution was more acceptable for preliminary analysis. Nevertheless, even the 60 x 90 resolution DEM was computationally time and space consuming. The 30 m resolution DEMs were not available for the whole study area, and would have been nearly unmanageable for analysis due to the amount of data for the extent of our study area.

Limitations occur when conducting broad scale analyses because resolution is coarse. For instance, there was a high proportion of sites on the lowland group. But, by looking at a map of the sites, it was clear that most of the sites were on lowland LTAs near the shorelines.

There were two main types of lowland landscapes: the lakeshore marshes and fens, and the more inhospitable interior bogs and swamps. Perhaps a distinction within the lowland group would have caught landform settlement preferences that fell through the LTA group filter. This study was intended to identify broad patterns of settlement representing collective decisions over many years, and to identify the limitations of each ecological level in discriminating settlement preferences. This example illustrates that single scale analyses, such as landscape level, may benefit from the addition of more detailed terrain overlays or other finer scale ecological units.

DISCUSSION

Prehistoric Settlements

Studies of distribution and changes in prehistoric settlement patterns in the Upper Great Lakes provided a basis from which to compare Woodland sites to landscape structure and to historic and modern era patterns (Anderton 1993; Franzen 1986, Martin 1985). For example, Franzen (1986) developed a descriptive locational model of prehistoric settlement on the Hiawatha National Forest based on ethnohistories, ethnographies, and natural resource data. In addition, Martin (1985) examined resource availability and technological repertoire along Great Lakes coastlines. Through combined geomorphic mapping and archaeological investigations, Anderton (1993) affirmed the assumption that Archaic coastal sites were located near spring spawning habitat, which would have included areas of shallow water in complex shoreline reaches.

Findings for the prehistoric and historic periods are generally in agreement with the prehistoric locational model by Franzen in 1986 of settlements on the Hiawatha. He identified sturgeon spawning areas, areas with open water, beaver, moose, deer, and plant resources, and bedrock outcrops for chert procurement as the primary factors for settlement location. Hammett (1992) described a similar exploitation pattern for historic era people in the Southeast which was based on "...a wide breadth of available resources. The coast provided the added resources of the sea to all their terrestrial resources.

Most major inland residences were also located near running water, and archaeological evidence indicated that this had been the pattern of occupation in the Southeast for at least the last 5000 years..." Similarly, in his paleoshoreline geoarchaeological study, Anderton (1993) noted that Archaic settings reflected seasonal use of low energy coastal environments which would have offered access to spring-spawning fish habitat, as well as other wetland and coastal resources. He also observed that certain sites appeared to be later re-occupations of paleoshorelines, representing Woodland sites with a focus on interior wetland resources. Familiarity with the specific locations in our database, and the combination of multi-scaled mapping indicated that combinations of the coastal and wetland resources would have been available at many of the bedrock and lowland settlement sites.

The prehistoric and historic sites at higher elevations in our datasets, on the other hand, were probably inland winter hunting camps or maple sugaring camps (Densmore 1974; Franzen 1986).

Historic Settlements

In most subsections and LTA groups the number of settlements decreased from prehistoric to historic times. We were unable to determine if this was a reflection of sampling discrepancies between the two time periods, or a reflection of the span of time covered by the Woodland Period (approximately 3000 years) versus those identified during the GLO survey (less than 20 years). However, based on estimates of populations from archaeological studies and historic reports, these data may represent an actual decline in native populations resulting from contact with European traders.

Epidemic outbreaks of smallpox and measles between 1630 and 1880 caused losses of one-third to one-half or more of the population afflicted in the Great Lakes region. Deaths occurred among the Ojibwa in upper Michigan from outbreaks of cholera in 1834 and 1849 (Tanner 1987).

Historic occupation patterns differed from those of prehistoric or present settlements. Site selection related to slope aspect, for example, was actually most similar between the prehistoric and present sites. Likewise, subsection 212He, along the Green Bay shoreline, was occupied in prehistoric and present periods more so than in the historic. These findings suggested that historic people were relocated from preferred site locations, or accepted a different kind of living site, in terms of location and terrain. Trade needs of that era such as fur trapping, may have necessitated a different setting than those selected in prehistoric or present times. Indian population in the northern Great Lakes region between 1768 and 1810 was approaching a stable distribution when the British fur trade was near peak production.

In 1768 there were 5000 Ojibwa in the region south of Lake Superior. At the same time, however, an unstable situation was rapidly developing to the south, and would soon move north. Americanization of the Upper Peninsula began with the founding of Fort Brady at Sault Ste. Marie in 1822 (Tanner 1987). Tanner's map of the Eastern Upper Peninsula in 1830 displayed sites of 25 Indian villages with an estimated population of 1500. Most land cessions in the same area occurred in 1836 and reserved Indian land was then located along Waiska Bay (west of Sault Ste. Marie), near Munising, Escanaba, and Nahma, and at an inland site at the approximate location of the inland sites. These historical events help explain the transitional nature of the historic period reflected in our settlement data.

Euro-American Settlement

Two Upper Straits French communities: Sault Ste. Marie at the falls of the St. Mary's river, established in 1668; and St. Ignace in 1671, were the first permanently settled Euro-American communities in Michigan with the establishment of their Catholic missions, and probably were still two of the only European communities by the time of the GLO survey (Santer 1977). The composite GLO maps (Trygg 1964, 1969) indicated areas of "private claims" and a mission in the vicinity of the 1408 ha site. Except for Escanaba, settled by 1830, most other areas of the Eastern Upper Peninsula were not settled until the late 1800s (Hudgins 1961).

Hudgins (1961) suggested that settlement progressed from lakeshores inward, however, when viewed over a longer time frame, we found abandonment of previously occupied inland sites, and greater concentrations of present development along lakeshores. The rate of population growth for each county from 1890 to 1960 was graphed. The majority of immigrants during this time period were from Canada, Finland, Sweden, and Ireland (Hudgins 1961).

Contemporary Land use and Socio-economic Trends

By comparing three time periods spanning over 3000 years we were able to approximate areas of growing human development density, or areas that were repeatedly occupied. Major areas of growth identified by this multi-scaled analysis included: the Green Bay shoreline (Menominee, Escanaba, Rapid River); the area between and including Sault Ste. Marie and St. Ignace; and shorelines along the peninsulas on the north shore of Lake Michigan. These findings were readily correlated with socioeconomic data for Michigan. For example, as of 1960, limited land cultivation occurred most frequently in Menominee, Delta, and Chippewa counties, near growing cities.

These counties also have the longest growing season of all the counties in the Upper Peninsula. Products such as hay, oats, potatoes, milk cows, and hens followed a similar pattern (Hudgins 1961; Senniger 1963). Airline routes, major highways and the Mackinac Bridge, completed in 1957, connect Menominee, Escanaba, and Sault Ste. Marie to the lower peninsula, Wisconsin, and Canada (Hudgins 1961). In addition, the shipping industry, with a major throughway at the locks connecting Lakes Superior and Huron, was portentous for growth to the Sault Ste. Marie area, and for Menominee and Escanaba, both secondary shipping ports (Santer 1977; Senniger 1963).

That contemporary development is occurring on many of the same or nearby sites that harbored prehistoric people is not surprising. Two hypotheses may be put forth to explain the correlation. The first is that similar sites are chosen because the sites have a combination of necessary and/or desirable biogeographic factors (e.g. a protected fish spawing embayment along with transportation ease). The second proposal relates to the concept that people tend to occupy places previously occupied, regardless of the site environment. Both of these hypotheses may be proposed for future research.

CONCLUSIONS

Modern development patterns in the Eastern Upper Peninsula are following a distribution that in many ways resembles the distribution of prehistoric occupations, though at a higher density than that of known prehistoric sites. A critical outcome is that present settlement growth, in many areas will obscure and threaten the integrity of the archaeological record. Historic settlements were fewer than either prehistoric or present, and were distributed differently as a result of the violent changes in the region around the mid-1800s. Settlement patterns were often displayed at one hierarchical level but not at others. The combination of several analysis scales operate as complimentary lenses for portraying changing settlement preferences.

Findings from this study will be valuable to planners and resource managers in estimating trends and modelling regional development scenarios. In addition, they will be instrumental for archaeologists in selecting sampling schemes that will more efficiently locate and protect historic and prehistoric sites. Most important, the analysis of settlement patterns, viewed among units of relatively uniform physiographic variation, can disclose mutual causation trends in landscape evolution and put forth prospectives on which to frame finer scale cultural/ecological research questions.

12

FROM HOLISTIC LANDSCAPE SYNTHESIS TO TRANSDISCIPLINARY LANDSCAPE MANAGEMENT

INTRODUCTION

This chapter discusses the changes of some important concepts since landscape became a subject of research and management. In particular, the validity of some early concepts, such as holism, is discussed in the new context of sustainability and landscape management and participatory spatial planning. It will be demonstrated how landscape research developed historically along different trajectories, which have not yet converged into an integrated transdisciplinary landscape science. The recent developments in landscape research are driven by factors different from the past and which in fact are trying to keep up with ongoing changes in environment, society and landscape. However, when it comes to management and planning all approaches are dealing with particular areas of land; a synthesis is needed and transdisciplinary cooperation is unavoidable.

Today, many authors and scholars have a pessimistic view of the future of the landscape in the sense that it is expected that many more qualities will become lost and that the future is very uncertain (Lowenthal 1997; Kolen and Lemaire 1999; Austad 2000; Pedroli 2000; Lörzing and Simon 2001; Fairclough and Rippon 2002; Lemaire 2002; Van Mansvelt and Pedroli 2003). Some even speak of a crisis of the landscape (Lemaire 2002). The increasing pace and scale of the actual changes make the landscape difficult to manage in a sustainable manner and cause alienation for the people who live in it (Lowenthal 1997; Palang, Alumäe and Mander 2000; Lörzing and Simon 2001; Lemaire 2002; Cosgrove 2003). The profound transformation of most landscapes since the Second World War is characterized by a loss of diversity, an increasing homogeneity and a fundamental break with the past (Antrop 1997; 2003a). New landscapes are created without history and without ecological links to the natural context.

The general trend is one of polarization between a very concentrated and intensely used land and vast areas of abandonment (Vos and Klijn 2000). In landscape-ecological research the term 'disturbance' became very popular and human activities are often seen as the causes of this disturbance (Antrop 2001). In Europe, one of the first warnings that "something was going wrong with our landscapes" was given in the First (Dob??íš) Assessment of Europe's Environment: "The richness and diversity of rural landscapes in Europe is a distinctive feature of the continent. There is probably nowhere else where the signs of human interaction with nature in landscape are so varied, contrasting and localised. Despite the immense scale of socio-economic changes that have accompanied this century's wave of industrialisation and urbanisation in many parts of Europe, much of this diversity remains, giving distinctive character to countries, regions and local areas" (Stanners and Bourdeau 1995).

Many initiatives have followed since, stressing different aspects of these endangered landscapes: natural (Wascher 2000), agricultural (Dramstad and Sogge 2003; Parris 2004), rural (Claval 2004), multifunctional land use (Brandt and Vejre 2004a; Mander and Antrop 2003), archaeological (Fairclough and Rippon 2002), humanistic and philosophical (Pedroli 2000). The problem has been addressed as a core topic in most of the recent conferences on the landscape, such as the meetings of the International Association for Landscape Ecology (IALE) (Mander and Palang 2004) and the Permanent European Conference on the Study of the Rural Landscape (PECSRL) (Palang et al. 2004). Workgroups and joint projects have boomed: Landscape Europe, Landscape Tomorrow, Ecoland, Europae Archaeologiae Consilium, European Landscape Character Assessment Initiative (ELCAI), the DELTA programme for strategic research, and others). In this series of events, the European Landscape Convention (Council of Europe 2000) is an important development as it triggered policymakers and also focused politicians in many countries on the landscape 'problem'. The number of initiatives and projects dealing with the landscape is still growing and not all are really concerted, although most pretend to be integrated and inter-or transdisciplinary (Tress et al. 2003; Tress, Tress and Fry 2005). Clear definitions are important and transdisciplinary work demands a common language and good communication (Zonneveld 1995; Antrop 2001; Olwig 2002; 2004; Tress et al. 2003).

THE EMERGENCE OF LANDSCAPE RESEARCH

Early landscape descriptions dealt with the geographical characteristics of foreign regions or countries. With the renaissance period in the 15th century the first paintings and pictures of landscapes appeared in the Western world (Olwig 2002; 2004; Lörzing and Simon 2001). Kolen and Lemaire (1999) see this as the emergence of a landscape conscience. The systematic exploration and description of landscapes begin with the Age of Discovery, characterized by a rapid development of cartography and the growing interests of naturalists.

Alexander von Humboldt defined landscape concisely: "Landschaft ist der Totalcharakter einer Erdgegend" ("Landscape is the total character of a region of the Earth", Zonneveld 1995). Although he was a pioneer in biogeography, physical geography and climatology, he always stressed in his writings the human and cultural aspects in the landscape and above all the aesthetical qualities, which he even considered mentally healing (Von Humboldt 1814).

The French geographer Vidal de la Blache (1922) had a more historical and humanistic approach to the landscape, although he used similar techniques of annotated sketches and his prose was not so different from von Humboldt's. The main difference is that he recognized the importance of the local community in organizing the landscape, which results in a regional differentiation that is not only based on natural conditions but also on culture and is manifest in settlement patterns and territories. Also here, landscape is seen as a holistic unity, which is well expressed in the proper names given to the 'pays' or landscapes. The description of landscapes and regions became synthetic 'tableaux' of idealistic landscape models (Claval 2004), and many regional monographs were published (Antrop 2000b). Although both von Humboldt and Vidal de la Blache implicitly include the perception of landscape and its aesthetic qualities in their work, they do not study landscape perception and aesthetics as such.

In Central Europe and Scandinavia geographers developed the theoretical concepts of the 'Landschaftskunde' (Passarge 1919-1921; Granö 1929) or 'Landschaftslehre' (Neef 1967). The geographer Oppel used the term landscape science for the first time in 1884 (Troll 1950). 'Landschaftskunde' had a specific meaning as defined by Johannes Gabriel Granö: "Die Aufgabe der Landschaftskunde ist es, die Landschaften und die in bezug auf ihre Landschaften einheitlichen Gebietsganzheiten, m.a.W. die an den Beobachter gebundenen Fernsichten und die an den Raum gebundenen, in bezug auf ihre Fernsichterscheinungen einheitlichen Gebiete zu untersuchen, zu beschreiben und zu deuten" ("The task of landscape science is to study, to describe and explain landscapes and the related perceived areas composed of viewsheds and the features and spatial units they contain") (Granö 1929, p. 57).

The landscape is seen as "die sichtbare Fernumgebung oder Fernsicht", the visual surroundings, but extends the concept to "alle sinnlich wahrnehmbaren Sondererscheinungen", i.e. all sensory experiences (Granö 1929, p. 56). Granö combined the approach of the landscape as a scientist and an artist, without integrating both (Jones 2003). Inspired by the new way of observing the landscape using aerial photography, Carl Troll developed a methodological basis for a holistic, an ecological and integrated approach to the landscape (Troll 1939; 1950; 1959; 1966; 1968). He called "Luftbildforschung ist zu einem sehr hohen Grade Landschaftsökologie" ("aerial photography is to a high degree landscape ecology") (Troll 1939). He introduced the term landscape ecology and called it an "Anschauungsweise", a way of looking at the subject and he considered it "a marriage between biology and geography" (Zonneveld 1995). The introduction of the term landscape ecology in this sense promoted a new holistic and interdisciplinary synthesis in landscape research and also reconfirmed that perception is an integral part of the concept of landscape. At the same time Richard Hartshorne in 'The Nature of Geography' (1939) considered the term landscape with its multiple semantic meanings to be too confusing and abandons it as the object of study in geography in favour of concepts as region and space (Muir 1999).

During the 1960s and 1970s a deductive, rationalistic and quantitative approach dominated the new orientation in geography. Based upon the optimistic development in economy and technology, the common focus of geographers upon landscape was lost and a divergence and specialization in geography started. Gradually geographers in Western Europe lost their interest in the landscape as a regional synthesis as well as in the visual appearance and aesthetics of the landscape as subject of study. Landscape synthesis was the concept used by few researchers during that tome (Drdoš 1983; Snacken and Antrop 1983; Richter and Schönfelder 1986; Moss and Milne 1999). Zonneveld (1980) defined the loss of the regional holistic synthesis of landscape research in geography clearly in a paper entitled 'Het gat in de geografie', 'the hole in geography'. He stimulated the landscape-ecological thinking, mainly from the German and Central-European schools, in the education of geography students in The Netherlands.

The lack of an interdisciplinary and holistic approach of landscape study became clear in the fast changing environment in crisis, with new challenges of natural, ecological, cultural and social issues. The changing attitude towards landscape was also expressed in many philosophical essays such as 'The Iconography of Landscape' (Cosgrove and Daniels 1988), 'Filosofie van het landschap' ('Philosophy of landscape') by Lemaire (1970) and 'De angst voor het nieuwe landschap' ('The fear of the new landscape') by Lörzing (1982), 'Past time, present place: Landscape and memory' (Lowenthal 1975) and 'Topophilia' (Tuan 1974). The first attempt to restore the interdisciplinary approach of landscape research was made by the Landscape Research Group, which was founded in 1967, and initiated the publication of the journal 'Landscape Research'. In 1972 the Working Group Landscape-Ecological Research (Werkgroep Landschapsecologisch Onderzoek, WLO) was created in The Netherlands (Zonneveld 2000), publishing the journal 'Landschap'. It attracted different kinds of geographers, biologists and ecologists, as well as social scientists and planners. Historical landscape geographers were less involved. These conducted important work in the field of settlement geography (Renes 1982), mapping and classification of historical cultural landscapes in The Netherlands (Vervloet 1984), and the emergence of historical ecology (Rackham 2000).

In 1981 the WLO organized an international and multidisciplinary meeting in Veldhoven to reflect upon the future of landscape research (Tjallingii and De Veer 1982). Basically, the broken link with the tradition of landscape ecology as defined by Troll in 1939 was re-established and led to the formal creation of landscape ecology. Also contacts between the Western-European approach with the schools of landscape science of the Central-European countries was renewed. Two new international journals were published focusing upon the landscape: Landscape and Urban Planning in 1986 and Landscape Ecology in 1987. In 1988 the International Association of Landscape Ecology (IALE) was created. Since then, landscape research expanded over many different disciplines. The earliest implementation of landscape-ecological concepts in planning and policymaking occurred in Slovakia (Ružicka and Miklos 1990).

Debates are still going on about the specificity of landscape ecology and the definition of landscape. Moss (1999, p. 138) gives a practical approach: "To me, landscape ecology is simply about the study of landscapes and of the need to derive understanding about landscapes in order to enhance our abilities to manage them more effectively. Landscape ecology is not the only field to focus on the landscape but it has emerged in the last few decades because, quite clearly, existing approaches that sought to address a whole range of landscape-scale environmental issues were proving to be inadequate". Indeed, many new environmental problems demand a better understanding of the functioning of landscape and ask for rapid solutions at the appropriate scale of research and actions focused upon planning (Golley and Bellot 1991; Bastian 2001; 2004; Opdam, Foppen and Vos 2001).

The new transdisciplinary approach is mainly found in the domain of the new emerging landscape ecology (Naveh and Lieberman 1994; Zonneveld 1995; Brandt 1999; Wiens and Moss 1999; Moss 2000; Naveh 2000; Bastian 2001; Opdam, Foppen and

Vos 2001; Tress et al. 2003). Thus as landscape research is widening new fundamental knowledge is needed as well as more practical applications. The integration is achieved by multiple exchanges of ideas and methods. When landscape ecology is seen as a transdisciplinary science it indicates that it is not just combining sciences (multidisciplinary), nor is it 'in between' sciences (interdisciplinary), rather it is situated above different sciences, trying to integrate them with a common perspective (Naveh and Lieberman 1994; Zonneveld 1995; Moss 1999). Nevertheless, activities parallel to the ones initiated by landscape ecologists continued to develop.

Historical geographers, cultural geographers and landscape architects followed their own courses. In 2000, the journal 'Landscapes' was created as an interdisciplinary forum for these disciplines.

Today, landscape science refers to the whole of the disciplines involved in landscape research, such as landscape ecology, historical and regional landscape geography, geo-archaeology, landscape architecture and environmental (perceptual) psychology (Antrop 2000b). However, landscape science nowadays is not synonymous with the earlier German concept 'Landschaftskunde', which would be literally translated in 'landscape science'. The concept of 'landscape' has multiple meanings and subtle differences in different languages exist. Simplistic translations have caused much confusion and unnecessary disputes. A careful use of the language of landscapes has been stressed many times (Lowenthal 1961; Brandt 1998; 1999; Moss and Milne 1999; Naveh 2000; Zonneveld 1995; 2000; Antrop 2001; Claval 2004; 2005; Olwig 2002; 2004).

The historical development of landscape research and the interaction between disciplines. Disciplines in square boxes are the actual ones that contributed actively to the development of landscape science. Printed in bold are disciplines (mainly geography and ecology) that defined basic concepts and methods for the actual landscape science. The concepts, techniques and methods that were important for this development are underlined. Some important scholars who stimulated the development are named in italics. The three arrows pointing to the outside right-hand side indicate that the umbrella of landscape science closes with the application of basic knowledge in the actual landscape management on the left.

THE MULTIPLE MEANINGS OF LANDSCAPE

In common language, the word landscape has multiple meanings, and these have been discussed since the beginning of the scientific study of landscape (Naveh and Lieberman 1994; Zonneveld 1995; Dover and Bunce 1998; Kolen and Lemaire 1999; Muir 1999; Antrop 2000b; Olwig 2002; Claval 2004). Landscape does not only refer to a complex phenomenon that can be described and analysed using objective scientific methods; it also refers to a subjective observation and experience and thus has a perceptive, aesthetical, artistic and existential meaning as well (Cosgrove and Daniels 1988; Lowenthal 1975; 1985). Consequently, the perceived landscape is immediately analysed by the observer, compared and evaluated with his/her knowledge and previous experiences.

Landscape was also used to refer to an organized, managed or even administrative territory as expressed by the French concept 'pays', the German 'Landschaft' and the Dutch 'landschap'. Finally, the term landscape is also used as a metaphor, such as in media landscape or political landscape. English became the main common language in this domain, so many specific terms had to be translated. Brandt (1998) pointed out how difficult it is to translate many fundamental concepts used in the English-dominated landscape ecology to other languages. Many, subtle shifts in meaning occur and make exact understanding and communication difficult. Zonneveld (1995) stressed the exact meaning and epistemology of words in his book 'Land Ecology'. Indeed, correct use of concepts is an indicator of the maturity and purity of a discipline. Many textbooks of landscape ecology contain extensive definitions.

Considering all these aspects, it is not surprising that the approaches to landscape are very broad and not always clearly defined. Most interest groups dealing with the same area see different landscapes. The word landscape has complex and multiple meanings that shift in response to the context in which they are used and by the background of the users. Also, the meaning of the term changes with changing environmental conditions. For example, the historical geographer Muir (1999) describes the following: landscape history and landscape heritage, the practice of landscape history, the structure and scenery approach, landscapes of the mind, landscape, politics and power, the evaluation of landscape, the symbolic landscape, the aesthetic approach, landscape and place. Claval (2004) sees the landscape as the synthesis that embraces geo-ecological relations, spatial patterns, scenic and aesthetical qualities and even social and cultural traditions.

Cosgrove (2003) distinguishes two approaches to the landscape: the ecological one and the semiotic one. Landscapes have a holistic, perceptive and dynamical character according to Antrop (2000b). The holistic character required a synthetic and transdisciplinary approach where natural and cultural aspects of our environment are structurally integrated (Naveh 1995; 2000). Berdoulay and Phipps (1985) recognize two organizational forms of landscape: the ecological one and the visual or scenic one. The perceptive character indicates that landscape is essentially a reality in the eye of the beholder and can only be studied fully when considering the observer as well. Olwig (2002; 2004) attributes a political power to the scenic aspect that is expressed symbolically by shaping particular characteristic landscapes. Perceptive qualities of the landscape are important in settlement geography (Antrop 1988) and geoarchaeology (Fry et al. 2004). The dynamic character of the landscape refers to processes and to their functioning (Forman and Godron 1986). Landscapes change continuously and build a unique history. The relationship between natural and cultural aspects varies in time and space and forms a basis of the regional component, the chorology and history.

The European Landscape Convention defines landscape as "an area, as perceived by people, whose character is the result of the action and interaction of natural and/or human factors" (Council of Europe 2000). This definition contains many important elements. First, landscape refers to an area, ergo a well-defined territory that is organized and managed. People perceive landscapes, and thus their scenic and aesthetic qualities for humans should be considered. Different landscapes exist because of their distinct character, which is the result of the continuous interaction of natural processes and human activities. History and ecology are essential factors in the structuring and understanding of landscapes. No reference is made to 'special' landscapes such as 'spectacular' or 'ordinary' ones, to rural, industrial or urban ones; all landscapes should be considered equally.

The European Landscape Convention also defines landscape protection as "actions to conserve and maintain the significant or characteristic features of a landscape" and landscape planning as "a strong forward-looking action to enhance, restore or create landscapes" (Council of Europe 2000). The goals are twofold: preserving the landscape diversity that gives regions and places their unique character, and to improve the quality of landscapes if necessary by restoration and creation of new landscapes of quality. However, no time horizon is given, and sustainability can only be assumed. The definition of landscape in the Landscape Convention returns to the original concept where landscape is seen as a unique synthesis between the natural and cultural characteristics of a region.

RETHINKING SOME FUNDAMENTAL CONCEPTS

Holism, Scale and Hierarchy, Context

Holism is a bio-philosophical theory that originated with the naturalists during the early 19th century. Although Alexander von Humboldt did not use the term holism, the 'Gestalt' idea is present in most of his work. Holism was also important for the Gestalt psychology and in particular as a theory to explain how human perception works. It is therefore not surprising that von Humboldt in his writings emphasizes the aesthetic and the emotional experience of the landscape simultaneously with the rational observation and measurement of physical variables such as temperature. A similar approach is found in J.G. Granö's work (Granö 2003).

The perceptive dimension in landscape is fundamental, as the landscape concept integrates an area of land with its appearance, the scenery. Interaction between mental concept and construct, perception of the environment and adapted behaviour leads to landscaping, i.e. shaping and organizing the land according to the needs of a (local) society and according to political, ethic and aesthetic values (Olwig 2002). As the needs and values change, landscape becomes a dynamic phenomenon that is in continuous transition. The holistic principle also implies that the significance of individual elements is not only determined by their intrinsic properties and values but is also determined by their context. This is a basis for coherence between elements forming an ensemble of greater complexity and often of a unique composition and history. Context values are important for ecology, aesthetics and heritage valuation (Antrop 2004a; 2004b).

According to holism, the landscape should be considered a complex whole that is more than the sum of its composing parts. This indicates that all elements in the spatial structure of the landscape are related to each other and form one complex system. Many new concepts were introduced to describe the properties of the metastructure of the landscape. Common ones are pattern, patches, mosaic, structure, holon, connectivity and connectedness and context. Very often they are used as synonyms or in a rather loosely defined way, causing confusion. For many researchers nowadays, holism is outdated and not taken seriously in scientific study (Antrop 2001).

This may be because holism is an abstract concept that is inherently difficult to apply, in particular when using so called objective quantitative methods. A useful additional concept that allows an easier practical application of holism is the concept of a holon, which was introduced as building block of the Total Human Ecosystem by Naveh and Lieberman (1994). They defined a holon as an open system that is part of a hierarchy. The ecodevice concept is closely related (Van Wirdum 1982). A holon is a more or less autonomous subsystem that functions and has emergent properties as a whole. It contains holons at a lower hierarchical level and regulates them, while higher hierarchical levels also regulate it.

As holons work more or less autonomously, we do not need to gather data of 'the whole is more than the sum of its composing parts' (which is impossible), but need only data of what is related to the intrinsic properties of the feature studied and its context that determines its major functioning. The context defines the scope of the study. The first step in the study should be the definition of the scale and context at which a feature should be studied. Hierarchical structuring of landscapes is a classic method in land evaluation of land classification (Webster and Beckett 1970; Vink 1980; Howard and Mitchell 1980; Pedroli 1983; Zonneveld 1995).

It offers an approach to break down the extreme complexity of landscape into more comprehensive entities that can more easily be handled, classified, studied and managed. A first and important task in all landscape studies is the definition of the scale at which the study will be done; a task that is mostly achieved indirectly by the definition of the study area, the scope of the study, the density of surveying and sampling and the resolution of mapping.

Scale became a core concept in landscape ecology (Forman and Godron 1986; Turner et al. 1989; Burel and Baudry 2003) as well as the information content in spatial patterns (Phipps 1981; 1984). Instead of a static property of a map, scale became an important explanatory variable in geostatistics.

The hierarchical and holistic principle for the landscape of La Geria in Lanzarote. Lanzarote is the most arid of the Canary Islands with only 152mm of annual rain and a continuous dry northeastern trade wind. La Geria is a valley between volcanoes filled with black ashes. In the 18th century a particular vineyard landscape of about 3000 ha was created here to produce the sweet malvasia wine.

Vines were planted in circular pits dug in the volcanic ashes about one metre deep. Along the northeastern side a semi-circular open dry stonewall was build around each pit to protect it from the drying wind without causing turbulence. All solar energy is concentrated in the pit and heats the volcanic ashes.

During the night, these cool rapidly and as they are highly hygroscopic, they extract moisture from the air, which is concentrated in the pit with the plant. Each of the pits functions as a holon, an ecodevice at the micro-scale. Repeating this thousands of times creates a new cultural landscape with a distinct identity at the regional scale. This example also illustrates the interaction between natural conditions and processes and human creativity.

THE QUANTIFICATION OF THE LANDSCAPE: LANDSCAPE METRICS

Landscape metrics or landscape indices were developed to describe and measure in a quantitative manner the 'transcendent' structural properties of landscapes. Many quantitative metrics are available today (Milne 1991; Turner and Gardner 1991; Turner et al. 1991; Farina 1998). The almost explosive development of landscape metrics was possible when spatial analysis using GIS and image processing became available and calculation facilities became powerful enough. A comprehensive overview is given by McGarigal and Marks (McGarigal and Marks 1995) with the FRAGSTATS software. Most of the landscape metrics focus upon the multi-scale spatial patterns formed by land cover for studying landscape-ecological functioning or assessing and monitoring diversity issues. The possibilities for making meaningful quantitative analysis of spatial patterns also depend upon the availability of geographical data, preferentially as maps.

Many landscape metrics remain abstract and are difficult to understand and to interpret. The method used for defining land units and describing their characteristics significantly determines the values of the landscape metrics obtained. They remain more as scientific tools than instruments of communication to be applied in planning and transdisciplinary communication. The wide variety of landscape metrics resulted in a discussion about their real significance and applicability in planning (Dramstad, Fjellstad and Fry 1998; Fry 1998; Wrbka 1998; Antrop and Van Eetvelde 2000; Botequilha Leitão and Ahern 2002; Parris 2004). In planning and policymaking synthetic and more transparent indicators are preferred, in particular when related to agricultural policy (Oñate et al. 2000; Dramstad and Sogge 2003; Wascher 2003; Parris 2004). Applications in the field of historical and human geography and geoarchaeology also offer new perspectives (Fry 2003; Van Eetvelde and Antrop 2005).

Palmer (2004) related the concepts from landscape ecology and the meaning of some landscape metrics to scenic properties and landscape perception. The extent of the study should be based upon the organism's perception of the landscape. For humans this means a definition of home range (action radius) and viewshed.

Patch size and shape metrics can be useful here. The grain is considered to be "the finest resolution at which an organism perceives spatial heterogeneity" (With 1994) or "the distance or area to which the species is sensitive in carrying out its functions" (Forman and Godron 1986, p. 125).

For humans this means that sample areas should be similar to property lots (Palmer 2004). Humans perceive a rather limited number of different categories (approximately 5 to 10). Naturalness is considered to be positively related to scenic value; homogeneity is positively related to coherence and coherence is positively related to scenic value and visible landscape complexity (Palmer 2004).

Palmer (2004) found that half of the variation of landscape preference valuation could be explained by landscape metrics describing the composition (area and percentage of land-use types) and configuration (such as homogeneity expressed by dominance, fragmentation, patch density or edge density, and diversity expressed by patch richness, evenness or entropy.

SPATIAL PLANNING, LANDSCAPE MANAGEMENT AND DESIGN

Future Landscapes: Learning to Deal with Uncertainty

The trends of the current landscape development in Europe are summarized by Vos and Klijn (2000) and confirmed by many national and regional studies. In general a polarization is going on between more and less intensive land use and the spatial concentration of activities. Important driving forces are transportation and mobility, urbanization and globalization (Antrop 1999; 2004a; 2000a). The result is a loss of landscape diversity and identity, of landscape character and sense of place.

Simultaneously, landscape heterogeneity and complexity change and thus also processes and functions in the landscape. Future development becomes difficult to predict and uncertain, in particular in the highly urbanized landscapes (Antrop 2004c; Tress et al. 2004).

Traditional landscapes become fragmented and are gradually replaced by new ones (Antrop 2003a; Van Eetvelde and Antrop 2004). Also, the pace and magnitude of the changes are increasing and full-scale up-to-date inventories become almost impossible to make. All this introduces new challenges for policymaking, research and management, in particular:

- the task to monitoring changes
- sustainable development and multifunctional uses of the landscape
- new forms of landscaping.

The implementation of landscape-ecological principles in policymaking and planning are still weak and of great concern for landscape researchers (Dale et al. 2000; Opdam, Foppen and Vos 2001; Bastian 2004). Multifunctionality becomes an important issue here, as well as the sustainability and the use of concepts such as natural and human capital and landscape services (Haines-Young 2000; Brandt and Vejre 2004b; Haines-Young and Potschin 2004). Also new techniques for communication and visualization of scenarios of future development become important (Tress and Tress 2003).

INTEGRATED MONITORING AND LANDSCAPE CHARACTER ASSESSMENT

Integrated monitoring at the landscape scale becomes urgent and vital in the rapid changing landscapes to obtain the significant data to base research and decisionmaking upon. Many (national) schemes have been proposed (Agger and Brandt 1984; Bunce 1984; Ihse 1996; Dramstad et al. 2001; Brandt et al. 2002; Slak and Lee 2003; De Blust and Van Olmen 2004; Howard, Petit and Bunce 2004). Important efforts are made in the field of modelling and predicting land-use and land-cover changes (Fresco et al. 1997; Lambin et al. 1997; Agarwal et al. 2002), very often in the context of climatic change and processes of globalization and without concern of cultural, aesthetical and spiritual values of the landscape. Decision-making at the landscape scale implies dealing with uncertainty (Antrop 2004c). Also, the rapid changes made the preservation of the natural and cultural heritage an important issue again, "before it is too late" (Fairclough and Rippon 2002) and renewed the discussion of landscape values (Palang and Fry 2003; Antrop 2003b; 2004b; Haines-Young and Potschin 2004). Also landscape character assessment and typology have become important in particular in Europe (Wascher 2000; Vervloet and Spek 2003; Mücher et al. 2003; Pinto-Correia, Cancela d'Abreu and Oliveira 2003; 2004).

Sustainable Landscapes, Heritage and Multifunctionality

The concept of sustainability in relation to landscapes can be approached in two different ways. First, it refers to the preservation of traditional techniques in rural or pastoral landscapes and of the land qualities and natural resources needed to sustain this over time. Second, it refers to a guiding principle for future landscaping. The first approach implies also the preservation of inherent landscape qualities and values related to biodiversity, habitats and water, and to the cultural heritage. This can only be achieved when the traditional practices, local customs and functions that created these values can be understood and maintained (Austad 2000). The culture heritage consists of material objects in their landscape context and immaterial values such as the sense of place, local traditions and customs (Fairclough and Rippon 2002; Palang and Fry 2003; Claval 2004; Antrop 2005).

The second perspective to sustainable landscapes offers principles for future development, in particular to sustain the rural economy in any possible way. Haines-Young (2000) proposed the concept of natural capital as a new paradigm for landscape ecology, in particular when applying landscape-ecological principles in sustainable development and landscape management in the countryside. Botequilha Leitão and Ahern (2002) suggest the use of landscape metrics to evaluate sustainable landscape planning. Potschin and Haines-Young (2003) proposed the use of the German concept of 'Leitbilder' as a framework for concretizing the visions in sustainable landscape management, and Bastian (2004) sees it as an expression of an integrative approach to the landscape. The concept raises many still unsolved questions. Sustainability should be applied to a wide diversity of landscapes: natural and cultural ones, traditional and contemporary ones, spectacular and ordinary ones. What has to be sustained in each of those?

In the perspective of the general polarization of the land-use intensity, it is clear that no single solution to sustainability exists.

The differences between the rural countryside and the urban development demand completely different approaches (European Commission 2001). Another important question is the definition of the scale and time horizon of sustainability one is aiming at. Also, as a planning guide, sustainability will be dependent upon the dynamics in a larger economical context and dependent of new technologies that might develop. Time or planning horizons in economical way are a priori defined according to the realization of the project, and they define the willingness to invest and engage for a predefined period of time.

However, no such scopes are clearly defined today for the 'soft', un-monetized landscape values. A striking example of the deficit here is the number of world heritage sites that are listed as threatened (UNESCO 2003).

The concept of multifunctional landscapes is related to sustainable development. It emerged only recently during the international conference on multifunctional landscapes held in Roskilde, Denmark, in October 2000 and resulted in a whole series of publications (Brandt and Vejre 2004a; Fry 2001; Mander and Antrop 2003; Naveh 2001). The reason for this sudden interest is the refocusing of landscape researchers, in particular in Europe, upon the new challenges in landscape management in the concentration areas of population and activities where landscapes become highly dynamic.

Integrating Landscape Architecture and Design

Landscape architecture evolved along a separate trajectory through history. Landscape architecture evolved from the garden architecture of palaces, in particular in Western Europe since the Renaissance (Olwig 2002; Lörzing and Simon 2001; Jellicoe and Jellicoe 1975). It was closely related to new urban planning as realized in the founding of new residential cities. Olwig (2002; 2004) showed how landscape design became an instrument of national governance and autocratic dominance. Many concepts and principles of the 'landscape gardening' were also applied in the modern urban and spatial planning, in particular during the post-war rebuilding and the creation of new towns. During the 1960s to 1980s less attention was given to the shaping of landscape as part of planning policy. Since the 1990s, landscape architects have become more involved in spatial planning, in particular for 'finishing' and 'integrating' new infrastructures and reshaping congested urban centres. Landscape architects are arguably primarily creative designers of new landscapes rather than researchers studying existing landscapes. Landscape-ecological principles are only gradually integrated into landscape architecture and planning (Steinitz 2001; Nassauer 1992; 1995; 1997) and a lot needs to be done, in particular in the basic training of the designers.

The New Synthesis Demands a Transdisciplinary Cooperation

The holistic character of landscape means that multiple approaches to the landscape are possible. Each perspective uses its own concepts and methods and these are not always similar or comparable. Three perspectives can be recognized. The natural sciences, where landscape ecology has a leading role, analyse the changing spatial structure and its functioning using mapping. Furthermore they tries to describe the holistic meta-reality of landscape as a complex system using quantitative indicators.

The human sciences offer several approaches. The historical geography and historical ecology are most closely related to the approach of natural sciences as they focus upon long-term changes and often use mapping as a tool to understand spatial patterns. The psychological, humanistic and semiotic approaches to the landscape focus upon the perception and experience of landscape and its existential meaning. The results are rarely mapped. Finally, some applied sciences developed their own approach to the landscape, such as landscape architecture and spatial planning. Their approach is partially analytical but most of all creative design. As such they try to remodel and shape landscapes for the future in an attempt to steer the spontaneous and rather chaotic changes that go on.

Profound reorganization of the land to adapt to changing societal needs is resulting in rapid and vast changes of the environment. In areas of increasingly intensive land use and concentration of people and activities, new landscapes must be designed to fit the multifunctional use of space in a more sustainable manner. Today, the concept of 'landscape' is in a profound transition. Landscape does not refer anymore solely to the traditional rural countryside or the spectacular nature. Multiple visions and values exist for the same landscape. A new holistic synthesis necessitates more elaborated transdisciplinary cooperation.

HUNTER-GATHERERS AND HUMAN EVOLUTION

Holocene foragers may indeed have filled a new niche in many places, but by analyzing how climate, flora, and fauna influence ethnographic foragers we can gain insights into how Pleisto-cene foragers should have differed from them. Contact with agriculturalists poses a greater challenge. When more powerful societies that practice horticulture, pastoralism, or intensive agriculture (all referred to here as agriculturalists) come in contact with foragers, they can rapidly alter forager subsistence and culture. The significance of such contact, however, varies considerably. Many Pygmies, such as the Mbuti, speak only the Bantu language of their farmer neighbors for whom they do some labor in exchange for food. On the other hand, the Hadza have had contact with agropastoralists for a long time but maintain their language. Their interaction has been limited mainly to the trade of meat and honey for iron and tobacco, and has altered the Hadza surprisingly little over the past century.

All we can do is perhaps give extra weight to those with less contact. We might also give extra weight to those in richer habitats, though the bias toward marginal habitats is not as great as usually assumed.

The greatest obstacle to using foragers as analogs of our ancient ancestors is that virtually all foragers in the ethnographic record have complex technology compared to premodern hominins. Human niches are defined to a large extent by technology, even among foragers. There are horsemounted, bow-and-arrow hunters of bison, harpoon hunters of walrus who travel in kayaks, salmon weir-fishers, and spear, blowgun, and net hunters.

Technology has allowed human foragers to occupy the full spectrum of terrestrial habitats, producing a wide range of cultural variation. Our ancestors must have varied considerably as well, so it seems pointless to reconstruct a single type of society. However, because we think the first modern humans arose from a particular population living in Africa 160,000 to 200,000 years ago, there would have been a limited range of variation in that particular population at that particular time. As modern humans spread to diverse habitats and developed diverse technologies, sociocultural variation increased.

Because the archeological record, with a few exceptions like Tasmania, reveals a trend toward increasing efficiency in technological evolution, we can assume that there was less efficiency (in extracting resources, for example) as we go further back in time. It is mainly the effects of increased productivity associated with technology that we need to subtract from ethnographic foragers when extrapolating to much earlier times.

Ironically, while foragers may be problematic analogs of humans in the past, they are certainly the most useful exemplars of humans in the present. To know which of our traits are derived and therefore require a separate explanation from traits we share with Pan, we must first characterize extant species. When we compare humans with other species with respect to traits like diet, group size, home range, mating system, or mortality rates, we need to measure these traits in foragers, not agricultural populations, if we are to understand the relevant selective forces that shaped modern humans. Here I provide an overview of foragers to put humans in primate perspective and to examine some basic relationships between social organization, and habitat, and technology so that their effects can be accounted for when modeling hominin behavioral evolution.

THE FORAGER SAMPLE

Although few hunter-gatherers or foragers exist today, they are well documented in the ethnographic record. Anthropologists have been eager to study them since they assumed foragers represented a lifestyle that existed everywhere before 10,000 years ago and characterized our ancestors into some ill-defined but remote past. In the past few decades, that assumption has been challenged on several grounds. Ethnographically described foragers may be a biased sample that only continued to exist because they occupied marginal habitats less coveted by agricultural people.

In addition, many foragers have been greatly influenced by their association with more powerful agricultural societies.

It has even been suggested that Holocene-foragers represent a new niche that appeared only with the climatic changes and faunal depletion at the end of the last major glaciation.

Despite these issues, the ethnographic record of foragers provides the only direct observations of human behavior in the absence of agriculture, and as such is invaluable for testing hypotheses about human behavioral evolution.

Comparative analyses face the problem of non-independent data (Galton's Problem) whether the units are species, cultures, or individuals.

When the phylogenetic relationships between species are known, they can be used to avoid Type 1 or Type 2 errors. Across human societies it is less clear whether genetic or linguistic relatedness would be preferable for constructing phylogenies because of genetic mixing and language replacement. The SCCS was assembled to avoid Galton's Problem, but consists of only 186 societies, of which about one-fifth are foragers. The results I present on kinship, marital residence, and the mating system are mostly from my previous analyses of the SCCS. Because here I am more concerned with the full range of variation and central tendencies, I use the complete forager sample even though there is no satisfactory phylogeny yet. Because I present some correlations, I have broken the sample into Old World versus New World. If relationships found in the whole sample also hold within each hemisphere, we can have more confidence that the relationship is real.

In general, evolutionary theory predicts that individual behavior will be adaptive and track ecological conditions, subject to constraints. Populations should also track local conditions since groups are seen as collections of fitness-maximizing individuals. This assumption should usually hold despite the fact that individuals are so often in conflict with each other. Two individuals might compete for scarce foods but both go about acquiring them with the same "most efficient" foraging strategy in a given environment. Thus, to the extent that certain aspects of social organization follow from foraging strategies, which in turn track ecology and technology, we should be able to explain relationships between habitat variation and social organization. To illustrate how ecology and technology shape social organization, consider the equestrian foragers. Several North American Indian tribes became specialists in hunting bison from horseback during the 1700s, after the Spaniards introduced the horse.

In the sample here, equestrian foragers have much larger home ranges and significantly larger local groups than do nonequestrian foragers. The horse increased hunting success rates and lowered travel costs, allowing the Plains Indians to specialize in hunting the high-yielding bison and follow them over vast distances in large groups.

When reconstructing periods before horse domestication with data from ethnographic foragers, we can ignore the equestrian foragers. If we are interested in the period before 30,000 years ago, we might exclude the arctic foragers because it was only during the last 30,000 years that very cold areas were occupied by modern sapiens.

Like the equestrian foragers, arctic foragers have larger home ranges than do others, but in their case it is due to the low density of foods, which requires covering large areas.

When we are interested in how human foragers respond to different environments, the full sample is most useful because it reveals how factors like very cold climates or horses alter traits like diet, group size, and home range. When we are interested in estimating parameters for modeling earlier periods, on the other hand, we need a more relevant sample. For this reason, I created a subsample by excluding those foragers who hunt from horseback and those where effective temperature (ET) is 13°C, which eliminates the arctic foragers.

HABITAT VARIATION AND DIET

Primary biomass (kg/m^2) is a measure of the productivity of a habitat calculated from climate data.

The plant matter it measures provides food for animals and so limits their numbers (secondary biomass). It excludes animal biomass and includes inedible plant parts, so it is not an ideal measure of food abundance for humans. For example, rainforests contain less human food per kilogram of primary biomass than savannas do, but because there are so many more kg/m^2 of primary biomass in rainforests there is usually more total food energy present.

Therefore, even though primary biomass is not ideal, it does provide a rough estimate of habitat quality, and since it has been calculated by Binford for most of the foragers, I use his data. In the sample here, primary biomass peaks around latitude 15° North and approaches zero at the poles. The northern and southern hemispheres are not mirror images since there is more land mass in the northern hemisphere. Mainly as a consequence of temperature effects on primary biomass, as Lee, 17 Kelly, 10 and Binford have previously reported, gathering falls off in colder temperatures at higher absolute latitudes, where there are fewer plants, whereas hunting and fishing increase. Males, therefore, contribute more to the diet at higher absolute latitudes (r .480, P .0005, n 158).

Male contribution to diet is also higher where fishing is more important (r .510, P .0005, n 155). The "marginal habitat criticism" noted earlier assumes that areas ideal for agriculture would have been ideal for foraging and that areas bad for farming would have also been bad for foraging. This is not necessarily the case. Some areas unsuitable for planting can be quite good for foraging. The map of foragers shows that the main bias in the sample is a geographic one due to history. The absence of foragers in the circum-Mediterranean and most of Eurasia is not because that huge area has rich arable soil and other continents do not, but because complex state societies arose earlier there and had incorporated or replaced all the foragers before ethnographies were written. In addition, a large fraction of the foragers in marginal (low primary biomass) habitats are those in Central Australia, who were not pushed there by agriculturalists since there were no agriculturalists on the continent prior to about 1800. The habitats with the lowest primary biomass are those of the arctic foragers. When the coldclimate foragers are excluded, a comparison reveals that forager habitats are not less productive than those of agriculturalists. Population density increases with primary biomass, as expected from basic ecological theory, but it levels off at a primary biomass of about 30 kg/m^2. The high-density foragers at primary biomass about 35 kg/m^2 are mostly the foragers of the northwest coast of North America. These complex foragers occupy habitats where salmon return to their freshwater spawning areas. Their rich supplies of fish, which these foragers preserve and consume year round, allow them to live in small territories in sedentary, socially stratified groups, often with frequent warfare and even slavery.

It is not habitat richness per se that explains the special qualities of the complex foragers but the seasonally abundant anadromous fish, which promote storage and investment in time-consuming technologies, such as weirs and smoke houses, with delayed returns.

Complex foragers may not have been rare toward the end of the Pleistocene, but for modeling earlier periods we should exclude them, and most are absent from the warm-climate sample.

GROUP SIZE, HOME RANGE, AND MOBILITY

Among foragers, there are typically three types of groups, the ethno-linguistic group (tribe), which may never assemble in one place; the residential or local group (camp or band); and the daily foraging party. Most foragers have several types of fission-fusion. First, every day the local group splits into smaller parties to forage (while some may remain in camp), then returns to camp to sleep. Foraging parties may also fission or fuse. One camp (local group) may fuse with other camps and, two months later, fission off again. Individuals also move back and forth between camps.

There is also residential mobility of a different sort when the whole local group moves the location of the camp to access new resource patches. The vast majority of foragers are quite mobile, especially in the warm-climate sample, where the median number of moves per year is 7. Dunbar argues that the number of individuals one interacts with on a regular basis is limited by the ability to keep track of social interactions.

On the basis of a correlation between group size and the size of the neocortex relative to the rest of the brain across anthropoid primates, he argues the group size of humans over some evolutionarily significant period must have been about 150.

This is certainly at odds with the size of the most salient group among foragers, the local group, which has a median population of 30. Could local groups have been closer to 150 if foragers occupied richer habitats in the past? The data do not support this, since there is no correlation between local group size and primary biomass. In richer habitats, where concentrated foods can support larger groups, we see a higher population density since the area occupied by the local group shrinks to minimize travel energy expenditure, but there is no increase in local group size. Across the full range of primary biomass, local groups show a conspicuous tendency toward a population of 30.

Hadza camps are at their smallest during the rainy season and their largest at the end of the dry season. This is because there are only so many permanent waterholes, which limits the number of possible camp locations.

When rain removes this constraint, the default is to disperse into smaller groups in which the Hadza say there is less bickering. This suggests a possible explanation for the remarkable consistency in local group population across forager habitats. Freerider problems probably set an upper limit on optimal, or equilibrium group size.

Beyond 30, conflicts between families may cause fissioning, and it is only when there are external constraints, like the prevalent warfare of some equestrian and complex foragers, that there are larger local groups. Home range is the area an individual occupies over a whole year. Because few home range data exist, I calculated a minimum estimate of home range, which I call local group area. The more hunting contributes to the diet, the larger is the local group area and the more frequently camps move, especially when we exclude the sedentary foragers. Only 25% of the foragers in the sample are sedentary, and most of these are the complex foragers of the Pacific Northwest. There is less mobility where fishing, rather than hunting or gathering, accounts for more of the diet. Thus, our ancestors may have been quite mobile before fishing was important. Home ranges are smaller in the warmclimate sample, where hunting contributes less to the diet, so they may have been somewhat smaller in the distant past if hunting was less important. Among warm-climate foragers, local group area (home range) decreases as primary biomass increases, since resources are more concentrated. However, once primary biomass reaches about 10 kg/m^2, local group area does not continue to decrease. This implies that most warm-climate Late Pleistocene foragers, even in rich habitats, probably had very large home ranges of about 175 km^2. This is larger than the city limits of Washington D.C., too large an area to defend as an exclusive territory, given the population density (median .18 km^2). However, residents might have repulsed outsiders upon encounter as do some foragers and some carnivores with large home ranges.

As primary biomass increases, the total area of the ethno-linguistic group shrinks but the population remains fairly stable (median 875 cold; 565 warm). Presumably, there are only so many people who can remain tied together in terms of language, as Birdsell argued with his magic number of 500.

Foragers tend to have friendly, co-equal relations with other local groups but not with other ethno-linguistic groups, at least when those others are agriculturalists. But what was it like before there were any agriculturalists? Australians are instructive since at the time of European contact all were foragers. While there were ethno-linguistic boundaries, they were not very sharp. There was some intermarriage between adjacent groups, but there was also some warfare.

Along with the evolution of language came both the means to tie together several local groups and the basis for dividing one cluster of local groups from others. Since the ethno-linguistic group is largely defined by language, there is no analog of ethnicity in other mammals, and the continuous movement of individuals back and forth between local groups within the ethno-linguistic group is probably unique to humans.

KINSHIP, MARITAL RESIDENCE, AND DISPERSAL

In most mammals, members of at least one sex leave their natal group at maturity, perhaps to avoid inbreeding. Among human foragers, couples often live with either the husband's family (virilocal) or the wife's family (uxorilocal). Virilocal residence is sometimes assumed to be analogous to male philopatry since males stay with their kin while females disperse.

This analogy is more appropriate for sedentary agriculturalists than for foragers who frequently move in and out of camps; still, marital residence does capture something about potential nepotistic aid. There has long been a debate about the most common pattern of marital residence among foragers.

The dominant view over the past few decades has been that foragers, like agricultural societies, are mostly virilocal.

An analysis of foragers that takes account of residence in the early and later years of marriage does not support this.

Most foragers are multilocal; the couple resides in camps with the wife's kin at times, the husband's kin at other times, occasionally with both, and sometimes with neither. Bilateral descent, tracing kin through both mother and father, is more prevalent among foragers (75%) than agriculturalists (25%), who are most often patrilineal.

One advantage of patrilineal descent is that it is easier to assemble a large group of reliable allies in times of warfare or disputes, since people know which patrilineal clan they belong to and where their loyalties lie. This is probably one reason that patrilineal descent dominates among agriculturalists.

However, an advantage of bilateral descent is that it maximizes the number of kin ties across camps, which facilitates visiting, finding mates, and moving to access seasonal resources.

Because bilateral descent, which facilitates multilocal residence, is possible only if one knows one's father as well as mother, multilocal residence likely evolved after pair bonding.

Language makes kin terms possible, makes it easier to keep track of all of one's kin, and makes it easier to invent fictive kin, so it facilitates movement between groups. It is difficult to say what dispersal was like before language, but if we exclude the predominantly virilocal agriculturalists, it is worth noting that we come to a different conclusion about the most likely ancestral state. Since both sexes disperse some in gibbons, orangutans, and gorillas, as well as in human foragers, the most parsimonious scenario has strict male philopatry among chimpanzees evolving only after the Pan-Homo split and before chimpanzees and bonobos separate. The same is true for exaggerated sexual swellings and large testes. Among the extant hominoids, only the two species of Pan live in promiscuous, multimale, multifemale groups with extremely male-biased philoparty.

THE SEXUAL DIVISION OF FORAGING LABOR AND THE MATING SYSTEM

Among chimpanzees 71% to 90% of hunting is done by males, so it is possible that early hominin males did more hunting than females as well. The specialization in hunting or scavenging by males in contrast to gathering by females may have begun much later, but perhaps as early as 2.5 mya.

The common view that males only hunt and females only gather is false, however. For example, Australian females do considerable hunting of small animals, while males in many foraging societies collect honey and gather fruit.

Among some net foragers, such as the Aka, men and women hunt together.

Still, the sexual division of foraging labor among humans stands in real contrast to the behavior of our ape cousins and other mammals. We know little about the actual mating system of foragers since there are no DNA paternity exclusion data. What I describe is only the observable "social" mating system, which probably underrepresents actual polyandrous mating. There is considerable variation, ranging from the Andaman Islands, where polygyny is said not to exist at all, to Australia, where 70% of married men among the Tiwi can be polygynous. Polygyny in Australia is related to gerontocracy; old men have much higher status (and more wives) than young men do. The percentage of married women with a cowife better captures the degree of reproductive skew, since a polygynous man may have 2 wives or 10, which will affect the number of bachelors. In the total forager sample this ranges from 0% to 90% (median 10). In the SCCS, about 10% of foraging societies are monogamous, 60% slightly polygynous (20% have cowives), and 30% generally polygynous (20% have cowives), making foragers as polygynous as agriculturalists.

Even where there is considerable polygyny, however, most marriages are monogamous and, except in Australia, two wives is often the maximum. In the SCCS, in societies where male contribution to diet is higher, there is less polygyny.

Hence, human foraging societies fit a pattern seen across mammalian species, where males are most likely to invest in offspring in socially monogamous species, while the vast majority of species in which males invest little or nothing are polygynous.

Among most foragers, men's foods, especially larger game, are shared widely outside the household.

This means that the best hunter's household may eat no better than the households of poor hunters, which forces us to ask what benefit a woman would gain from marrying a good hunter. Some have suggested that a man's hunting success signals something about his quality other than his provisioning value to his own household.

While that may be true, I found that Hadza men who were not stepfathers brought more calories back to camp when their wives were nursing an infant and had decreased productivity.

Hadza men, therefore, appear to compensate for their nursing wives' lower productivity, which suggests that pair bonds may be maintained by the benefits of a sexual division of labor within households. Age at weaning ranges from 1 to 4.5 years (median 2.5), which is considerably younger than the 4.8 years for chimpanzees and bonobos (Tables 2 and 3). Infant mortality ranges from 10% to 46% (median 21) and juvenile mortality from 20% to 61% (median 45), slightly lower than the rate for chimpanzees.

Total fertility rate ranges from 0.81-8.5 (median 5.5). As I have reported elsewhere, higher mean male contribution to diet across foraging societies predicts a younger age at weaning, a higher total fertility rate, and greater female reproductive success, but not a lower offspring mortality rate. Even when infant and juvenile mortality is high, where men contribute more to the diet a greater absolute number of offspring survive to the age of 15 years. This implies that women may use food from men mainly to speed up their rate of reproduction.

CENTRAL PLACES, FOOD SHARING, AND EGALITARIANISM

Central place foraging is typical of most human foragers, as it is among many birds and social carnivores (Table 3). Because there are species that return to central places without food (among them hamadryas, Papio hamadryas, and spider monkeys, Ateles geoffroyi), a better term for those who do take food back might be central-place provisioners.

One reason for having a central place is to leave vulnerable or burdensome young in a safe place. Hadza mothers take their nursing infants with them when they go foraging but leave weanlings in camp because they are too young to keep up and too big to carry. Someone needs to stay in camp to look out for them and so food must be taken back. Perhaps this sets a lower limit on local group size among foragers; with less than 25 people and most adults out foraging, there might often be too few people available to baby-sit. As Glynn Issac long ago suggested and Nicholas Blurton Jones later modeled, food sharing among foragers probably began with tolerated scrounging not so different from the meat sharing seen in chimpanzees, who sometimes hold out their hands in a begging gesture, often turn their backs to beggars, and sometimes grab meat away from others.

Resisting the demands of several other hungry adults could be especially dangerous in a species capable of forming temporary coalitions, as presumably our ancestors were. Such coalitions can create what Boehm calls a reverse dominance hierarchy, resulting in egalitarianism. Food sharing would have lowered variation in daily food consumption, which should have lowered mortality rates and altered life-history patterns.

Central-place provisioning increases food sharing, since taking food back to camp creates more opportunities for scrounging. Because central-place provisioning makes food sharing and egalitarianism more likely, if central places can be detected archeologically, their earliest appearance might set an upper limit on the antiquity of egalitarianism.

A HIGHLY VARIALE AND UNUSUAL PRIMATE SPECIES

They also reveal many significant differences between Old World and New World foragers. Less of the diet comes from gathering and more from hunting and fishing in the New World, and male contribution to diet is higher. Ethnolinguistic population and area are larger, as is the local group area (home range). There are fewer moves per year, and the age at weaning is younger.

There are also differences between the cold-and warm-climate samples. Many of the differences mirror those between New and Old World. This is partly due to the fact that there are so many northern foragers in the New World, with the mean absolute latitude being 44° and ET 13°C, compared to 15° absolute latitude and ET of 20°C for the Old World. The latitude difference between New and Old World is present even in the warm-climate sample (NW 32°, 16°C; OW 17°, 19°C). When latitude is controlled, many of the differences between New World and Old World disappear, though fishing is still more important and total area of the ethnolinguistic group is still larger in the New World.

In comparative perspective, humans are so behaviorally diverse that societies are more like different species than different populations of one species, although there are some common features shared by most. In some of their widely shared features, human foragers resemble social carnivores more than they do our closest primate relatives. Like wild dogs, foragers have long day ranges, large home ranges, central places, provisioning, food sharing, and bisexual dispersal (Table 3). Many of these traits are probably due to technology that allows us to acquire large amounts of food that can be trans-ported and is consequently more likely to be transferred. This does not have to be meat; any food, such as honey, that is taken back to a central place in large quantities is more likely to provoke demand-sharing.

While technology has intensified greatly in just the past 100,000 years, if the hominin divergence from Pan entailed a shift to high-quality but difficult-to-acquire foods dispersed in a more open habitat with increased threat of predation, some increase in technology may have been selected for from the very beginning.

TECHNOLOGICAL CHANGE

The behavioral evolution of our ancestors must be reconstructed trait by trait, with the time period and area specified.

For certain traits, the ethnographic foragers are relevant for only the past few thousand years; for other traits they may be informative about hominins millions of years ago. Table 4 shows some of the important technologies and their probable effects. There is not space here to do justice to the complexity and ambiguity in the archeological record, so I have chosen to show both the oldest solid dates, as well as the oldest proposed dates, regardless of how controversial.

Many foragers acquire iron through trade and use it to make arrowheads. Iron first appeared about 2000 to 1500 B.C. in Western Asia, and by 600 B.C. had spread far and wide in the Old World, though not to Australia. Iron's superiority for durable, effective tools and weapons explains why it is so highly valued and spread so rapidly. Since most contemporary foragers use iron rather than stone or bone points, they probably spend less time on tool manufacture per tool than did Pleistocene foragers and so have more time to allocate to other activities.

Other fairly recent hunting technologies include traps, snares, and nets. As noted, net hunting can influence the sexual division of labor because males and females more often hunt together.

The possession of iron tools, cooking pots, factory-made water buckets, and cloth make contemporary foragers different from Pleistocene foragers, but do not make them so different as to be uninformative.

Because the quality of their arrow poison can vary, the Hadza often hit animals that do not die. Hunting large game without poison would have been more difficult. Poison use has received little attention archeologically and will not be easy to document. Hadza men with bows and poisoned arrows usually hunt alone because the challenge is getting close enough to prey to get a shot before being detected, and two people only increase the chance of detection. Bows may have caused a switch from group hunting to more individual hunting, or at least hunting in smaller groups.

13

Cultural Persistence and Environmental Change

Ethnohistorians ascribe the persistence of distinct ethnic groups in colonial Mexico to "the simple degree of contact, measured in distance, frequency, or hours spent" (Lockhart 1992:4-5). It is true that groups distant from the center of colonial society, and with little to attract colonial entrepreneurs, tended to remain ethnically and culturally distinct, whereas groups in the central regions exhibited the greatest degree of change. But there are exceptions to this rule: one such is the Otomí of the Valle del Mezquital, located a scant sixty miles north of Mexico City. The Otomí have been in continuous and close contact with first Spaniards and then Mexicans for nearly five hundred years, and yet they have remained linguistically, culturally, and ethnically distinct from the neighboring ladinos. Indeed, Charles Gibson (1967:10) notes that at contact and throughout the colonial era (1521-1810) they spoke the only non-Nahuatl-based language in the Valley of Mexico-and they continue to do so. The region they inhabit, the Valle del Mezquital, has also achieved a distinction of sorts in the twentieth century: as the archetype of the barren regions of Mexico, and as a vast, informal treatment plant for the waste waters of Mexico City.

The Valle del Mezquital comprises the catchment area and drainage basin of the Tula River, and in the 1930s an irrigation district was created that channeled the flood waters that regularly inundated Mexico City into the Tula River system. These waters were used to irrigate lands in the Tula River valley, and as the effluent from the city increased in volume the irrigation district grew in importance and extent. Despite its current status as a major source of vegetables for the city's markets, however, the Valle del Mezquital continues to be perceived as a traditional, backward, and poor "indigenous" region by scholars and lay people alike. Indeed, it is an indication of the power of this perception to shape policy that the chemicalladen waters of Mexico City's effluent could be thought to "improve" the soils of the region.

The perceived poverty of the Valle del Mezquital is invariably identified with the land-management systems of the Otomí. Geographer Kirsten Johnson's study (1977) comparing the attitudes of the Otomí toward the land and its use with the technocratic approaches of government experts sent to prepare Otomí villagers for the expansion of irrigation system #3 demonstrates attitudes toward this region and toward the Otomí that persist until the present. Not too surprisingly, perhaps, her study demonstrates profound differences in goals and approaches between the two groups: the technocratic approach to land use in Mexico's arid regions is based on the idea that scientific knowledge rationally applied could correct the perceived poverty of both the soils and the people of these regions; the Otomí approach is summed up in the saying "Do as the land bids." For the technocrats, the poverty of a region such as the Valle del Mezquital is a consequence of outmoded technology and a lack of access to scientific knowledge; they argue that formal education would teach new ways to use the land, which would provide the means to correct the poverty of both the Otomí and their region (Johnson 1977:161). For the Otomí, their poverty is a consequence of their lack of access to water-ultimately a lack of power rather than education. They argue that access to irrigation water would allow them to grow more and different types of plants. While they appreciate modern material technology-tractors, for examplethey express no need to change radically their approach to the land or their land-management systems (Johnson 1977:147-150).

Drawing on her own ethnographic research and the secondary literature dealing with the Otomí, Kirsten Johnson developed a synthesis of Otomí land use and attitudes to the land. (Since her study was completed, irrigation system #3 has spread to include extensive areas of the Valle del Mezquital; the practices she discusses are still carried out in areas where the national irrigation system does not reach.) The Otomí apply what might be termed an "ecosystem" approach to agriculture: they use their detailed knowledge of local conditions to select niches for particular species according to the type of soil and probability of rain and/or floodwater. Yet they do not simply rely on the bounty or lack thereof from the land. Rather, they also practice various methods to increase the productivity of the soils: canal irrigation, hillside terracing, floodwater dams, and wetlands management (drainage mostly) are used to obtain optimal soil/water ratios; rotting vegetation and animal manure are used to improve soil structure and fertility (Johnson 1977:110-147).

The technocrats interviewed by Johnson viewed these techniques as wholly inadequate, however, since they provide only a bare subsistence for the Otomí themselves and contribute little or nothing to the national economy. The modern landscape is assumed-certainly by the technocrats, but also by the majority of social scientists-to reflect the continuity of ecosystems as well as culture from the pre-Hispanic era (Johnson 1977:161).2 But when the history of the Valle del Mezquital is examined more closely, one finds that far from reflecting a traditional indigenous past, the modern environments and landscapes of this region reflect profound environmental changes set in train by the Spanish invasion; and that although certain Otomí agricultural techniques are almost exact replicas of those found in the archaeological record-for example, their methods for managing soil-water regimes (Sanders, Parsons, and Santley 1979)-they are applied to very different environmental conditions.

In the pages that follow I will explore the hypothesis that the transformation of the environment associated with the introduction of sheep by the Spaniards in the sixteenth century brought about the effective marginalization of the Otomí, so that while they were not distant from the center of Spanish society they were structurally isolated and hence were forced/allowed to remain distinct. I will argue that the mode by which sheep-grazing was introduced into the Valle del Mezquital allowed for its adoption by the Otomí, while yet allowing for the retention of traditional land-management systems in the same space.

THE INTRODUCTION OF SHEEP

The Valle del Mezquital was, at the time of initial contact between the Otomí and the Spaniards (ca. 1524), a densely populated agricultural region where extensive areas of land had been cleared of trees to make way for agricultural fields. Despite the extent of forest clearance, however, sufficient water was generated within the catchment area of the Valle del Mezquital to sustain the flow of springs and streams and to support quite extensive irrigated systems; only the hillside towns and villages in the very arid northeast were without irrigated fields. The extent of irrigation, together with the presence of perennial streams and springs, indicates a healthy catchment area until the middle of the sixteenth century. By the end of that century, however, the catchment area had deteriorated, and there was no longer sufficient water available to support both Otomí agriculture and growing Spanish enterprises such as agricultural holdings, flour mills, and mines (Melville 1994b: 31-39).

The deterioration of the catchment area of the Valle del Mezquital over the last quarter of the sixteenth century resulted from the expansion of sheep-grazing into this region following the Spanish defeat of the Mexica in 1521. As the Spaniards expanded into the Valle del Mezquital, regional production shifted away from intensive indigenous irrigation agriculture to small-scale intensive sheep-grazing that was characterized by high densities of animals (reaching ca. 785 head of sheep/km2 in the Tula River Valley in the 1570s) and large numbers of flock owners who had title to (or squatted on) small areas of land.4 The initially abundant vegetation and the custom of grazing in common obviated the need to own large areas of land in order to secure forage for flocks that numbered ten to fifteen thousand (and at times twenty thousand) head in the 1570s.5 By the 1570s sheep-grazing dominated the region.

At the same time, the Spaniards accelerated wood cutting for mining and for lime and charcoal manufacture, cleared forests for grazing, and fired the range for pasture regeneration; by the 1580s, the hills were cleared of trees. With increased soil temperatures and the development of more arid microenvironments, the vegetative cover shifted from a complex mosaic of cultivated species such as maize, beans, squash, tomatoes, and maguey; shrubs such as mesquite (growing as singlestemmed, massive trees); isolated dicots such as willows, cedars, and native cherries; forests of pines and oaks; and open woodlands of mixed species, to a more homogeneous desert scrub dominated by small, multistemmed mesquite. Water was no longer detained for sufficient time to replenish groundwater; instead, it flowed rapidly over the surface, carrying the soils down from the high, steep-sided hills onto the lower slopes. By the 1590s extensive gullying had begun to cut down through slope wash and the underlying soils; at the same time, water tables dropped, and springs failed (Melville 1994b:39-59).

The transformation of the physical environment of the Valle del Mezquital enabled the Spaniards to gain control over this region. By the end of the sixteenth century, two-thirds of the surface area had been transferred into the Spanish system of land tenure; the Otomí villages and their much-reduced agricultural lands were surrounded by extensively grazed latifundia; and the Otomí were marginalized within the new political economy (Melville 1994b:116-50).

This collapse of indigenous agroforestry was not simply the result of Spanish expansion, however, at the expense of indigenous communities, nor was it a straight-forward case of ranching competing with agriculture, for many Otomí adopted sheep-grazing on a large scale themselves during the sixteenth century. The sheep stations that surrounded Otomí villages in the sixteenth century may well have been the property of the villages themselves, or of Otomí nobles; and the sheep that transformed the biological regime were owned by Otomí as well as Spaniards. The Otomí had become agents of the profound transformation occurring in their landscape.

THE ADOPTION OF SHEEP-GRAZING BY THE OTOMÍ

Otomí were grazing sheep as early as the 1550s, apparently without license to do so. In the 1560s, Viceroy Velasco began to grant licenses for sheep-grazing to individual Otomí nobles and to Otomí communities; during the six-year period 1560-1565 they received 78.9% of the mercedes for estancias de ganado menor (grants of land for corrals and shepherds' huts, and license to graze sheep or goats in common in the surrounding area), and by 1565 they owned 35.8% (99) of the total 276 sheep stations in the region. In the years 1566-1569, 5 of 11 grants went to Otomí, almost as many as to Spaniards; and in 1570-1579 Otomí received 26.5 of 44.5 grants-considerably more than the Spaniards. And although Spaniards acquired more stations by squatting during these years, the Otomí still held slightly more than one-third of the lands given over to sheep-grazing by 1579.

The acquisition of sheep stations by the Otomí declined in the decade of the 1580s, and they received only 18.5% (13) of a total of 70 grants. But in the 1590s, Otomí nobles and villages again received more grants than the Spaniards: 69.8% (94) of the 154.5 grants awarded. By 1600, 37% of the grazing lands were in Otomí hands (Melville 1994b:116-150).

Until the end of the century, when severe restrictions were placed on Indian participation in the pastoral sector of the political economy of New Spain, many Otomí seem to have grazed animals far in excess of the legal maximum of 2,000 head per station-as did at least one "mulatto" and many Spanish pastoralists (Melville 1994b:49-54).

The process of land acquisition through the Spanish system of land tenure does not negate the fact of the eventual economic marginalization of the Otomí, however: by the seventeenth century Otomí villagers were transformed from the dominant economic force in the Valle del Mezquital into marginalized peasants.

Nor does it negate the fact that Otomí communities had no room for future expansion: so much land had been transferred into the Spanish system of land tenure by 1600 that, despite the demographic collapse of the Otomí population, which had reached 90-92 percent by this date, there was insufficient land for the fondo legal (townsite to which every community was entitled) of the villages in the fertile southern half of the region (Melville 1994b:146-150). But it does put a rather different complexion on Otomí survival and persistence-making it difficult, at least at first sight, to argue structural isolation or cultural involution as their means of cultural survival.

PARALLEL SYSTEMS OF RESOURCE EXPLOITATION

The enthusiastic adoption of sheep-grazing by the Otomí, the profound transformation of the vegetative and soil cover of the Valle del Mezquital under sheep-grazing and the consequent loss of traditional resources, the proximity of the region to the center of Spanish power, and the intensity of contact over several centuries are the same forces that in many other regions transformed "Indians" into "Ladinos." That transformation invites questions with regard to the Otomí. Specifically, how did they remain so distinct? How were they able to adopt a cultural element characteristic of the conquerors-one that was, moreover, a major sector of the evolving colonial political economy-and yet not become ladinos? How were they able to maintain traditional systems of land management when their physical world was transformed literally under their feet?

In reply to these questions, I would propose a mechanism that may have enabled the Otomí of the Valle del Mezquital to adopt pastoralism while at the same time maintaining agricultural practices and approaches to the land that were distinct from Spanish or mestizo systems of land management. This mechanism concerns the formation of ranching as a parallel system of production outside Indian village/ community production. Historically, environmental degradation brought about Otomí marginalization, destroyed the basis of intensive pastoralism itself, and undermined irrigation agriculture using internally generated water.

The combination of political and economic marginalization on the one hand, and physical isolation in a transformed landscape on the other, I suggest, produced a situation that almost guaranteed Otomí ethnic persistence. It should be noted, however, that I am not suggesting that the Otomí remained the same, nor that they did not undergo change; simply, that they remained distinct from Spaniards and mestizos, and, later, Mexicans. I originally constructed the model of parallel systems of resource exploitation as a means to conceptualize the processes by which Spanish pastoralists and their animals were able to expand into densely populated agricultural regions, such as the Valle del Mezquital, without the use of organized force-though with violence, as will be seen. I suggested that the processes by which pastoralism evolved in New Spain were essentially different from those by which Spanish colonial agriculture evolved. Whereas Spanish agriculture was, from the first, embedded in Indian systems of land tenure and use together with village systems of production, pastoralism evolved "outside"-or better, parallel to-village production (Melville 1990).

The initial development of Spanish agriculture occurred within the context of the encomienda system set up immediately following the Spanish defeat of the Triple Alliance, as a system of rewards given to individual conquistadores. As an encomendero, a conquistador was to be supplied with labor and tribute from a specified community of Indians in return for the defense of their customary rights after the collapse of any centralized indigenous authority, within the extremely unsettled conditions of the early conquest era when Spanish authority was insecure. ("Conquest" is here equated with the process by which the colonial regime evolved.)

While the encomienda system clearly served to siphon off tribute to the conquerors, it also functioned as a decentralized system of government. And because it resembled, in broad terms, indigenous tribute systems, it was imposed with remarkably little difficulty.

Although encomenderos had a virtual monopoly on production within the Spanish world up to 1550, primarily by virtue of their access to indigenous labor, they were ultimately dependent on indigenous land-use and land-tenure systems for their agricultural enterprises. They could state the tribute they required and the crops they wanted cultivated, and they could direct their laborers into a variety of entrepreneurial pursuits; but they did not have rights to the lands encompassed by their encomienda communities.9 Crops were grown on Indian lands using Indian labor, very often according to indigenous management systems and schedules. We find wheat being cultivated in the mounds typical of maize agriculture in the early years, for example (Gibson 1964:322).

And later, when it became clear that wheat production on a commercial scale competed with Indian work schedules, most especially maize cultivation, and that the best grains for bread could not be successfully grown during the summer rainy season, the Spaniards changed their scheduling: from spring sowing and fall harvest of rain-fed wheat, to fall sowing and spring harvest of irrigated wheat. As Arnold Bauer (1986) has demonstrated, the shift to a spring harvest meant that threshing was postponed to the following fall, when traditional methods of threshing with the trilla, or with large numbers of mares on a flat area of hardened ground, were not threatened by early rains and did not use animals needed for preparing the ground for maize.

Encomenderos bought or otherwise acquired land (by squatting, grants, gifts, or force) within their encomiendas (Gibson 1964:272-274)-thereby appearing to remove their enterprises from dependence on Indian production systems. But a case can still be made for the dependence of the new agricultural enterprises on the villages by examining developments following the imposition of the repartimiento in 1550. In a move to reduce the power of the encomenderos the Crown took to itself the right to Indian labor and set up government-controlled work gangs: repartimiento in New Spain, mita in the Andes (Gibson 1964:224-236). The shift to the repartimiento system of labor recruitment took the control of Indian labor out of the hands of the encomenderos and made Indian labor available to non-encomenderos, but it did not eliminate the Spanish agriculturalists' dependence on Indian villages; rather, it consolidated this dependence, producing a land-labor system that was carried into the era of the hacienda. Indeed, not only is management of a grain or mixed hacienda inconceivable without Indian villages to supply the work gangs needed to weed and harvest, but the relations between haciendas and Indian villages are treated as a defining characteristic of the hacienda system (Van Young 1983; Gibson 1964:292).

In contrast to the development of Spanish agriculture, the introduction and expansion of pastoralism were not dependent on Indian labor, land tenure, or landmanagement systems. Pastoralists did not need title to a specific area of land, since grass was a "fruit of nature" in Spanish custom, and animals could graze wherever grass grew (Vassberg 1984:6; Chevalier 1975:12). Nor did they have to rely on Indian labor to care for their animals: they seem often to have relied on black slaves in the early years.10 There was really no reason why anyone who could afford the cost of a few breeding animals and of someone to care for them could not set up a herd, request a grazing license, and be in business. Thus pastoralism evolved, in structural and organizational terms, in parallel to the village-encomienda system of production; it was not embedded within it.

That most early pastoralists were encomenderos, however, is not surprising, since they had the means for capital accumulation and thus the means to acquire breeding stock and develop herds (Miranda 1965; McLeod 1973:46). Nor is it surprising that they often did not bother to obtain licenses, but simply squatted on the lands they needed for corrals and shepherds' huts and grazed their animals in common on the surrounding lands (Gibson 1964:275; Simpson 1952:6; Chevalier 1975: 131). But it is interesting to note that in the Valle del Mezquital, encomenderos grazed their animals in others' encomiendas in the early decades, and rarely in their own: of the fourteen local encomenderos grazing herds of sheep and cattle in the Valle del Mezquital during the 1530s, only three grazed their animals in their own encomienda (Melville 1994b:128-129 n. 25).

Perhaps it was simple pragmatism that kept the encomenderos from grazing in their own encomienda lands. Relations between the herdsmen and the villagers was notoriously violent; indeed, violence seems to have functioned as a means to expand into this densely populated agricultural region and to gain access to forage.

In such circumstances, it would make more sense to annoy someone else's tributaries rather than one's own. But the independence of the encomendero-pastoralists from their own village-encomiendas in the Valle del Mezquital nicely illustrates my point that pastoralism was not structurally and organizationally embedded in the village-encomienda system of production.

As well as serving as a conceptual model for the Spanish invasion of densely populated regions, the model of pastoralism as a parallel system of resource exploitation also serves to clarify the development of the striking differences between the labor regimes of grain and livestock haciendas, and further illustrates the development of pastoralism in New Spain. The marked independence of laborers on livestock haciendas is often ascribed to the low labor needs and less-intensive landmanagement systems of traditional pastoralism (J. Riley 1984:261-262; Zeitlin 1989; Gutiérrez-Brockington 1989:103). I suggest, however, that such differences were not simply the result of the different labor needs of pastoralism and agriculture; rather, they arose in the formative pre-hacienda era as a result of the different developmental processes by which grazing and Spanish agriculture evolved. I base my argument on differences between the labor regimes of agriculture and pastoralism that appeared at an early stage in the Valle del Mezquital. Various categories of wage laborers (sirvientes, "gente," naboríos, gañanes) were resident on the smallscale agricultural and pastoral units of production prior to the appearance of the hacienda in this region.13 But while Indian sheep-shearers living in villages hired themselves out to estancia (sheep station) owners, agricultural specialists did not live in villages and hire themselves out to labor (agricultural units of ca. 200 ha) owners; rather, they lived on the labores in a manner reminiscent of the later hacienda system.

The structural and organizational separation of pastoralism from village production provided Otomí commoners with an opportunity to learn new skills and earn money outside the community-and perhaps to escape increasingly onerous obligations (Zeitlin 1989:47; Hoekstra 1993). It also provided a means for the Otomí elite, and for entire Otomí communities, to engage actively in the new political economy (Melville 1994a). The benefits of raising domestic livestock were many: it was a relatively easy and secure way to obtain needed items such as fat, meat, fiber, traction, manure, and leather. The Otomí evidently preferred sheep over cattle. Several factors shaped their choice of these animals. First, sheep have one great advantage over the other domesticates brought by the Spaniards: their wool; wool is a much warmer fiber than cotton, eminently suitable for the cool highlands, and this would probably have influenced Otomí acceptance of these animals. Second, sheep thrived particularly well in the cool highlands. Third, mutton was preferred over beef in the sixteenth century (Chevalier 1975:142-143). Finally, the development of the Otomí as sheepherders rather than as, say, cattle ranchers was probably shaped by governmental decrees: ganado mayor (cattle and horses) were officially expelled from the central highland regions by viceregal order in the 1550s (except those animals destined for use on agricultural lands), and the region was thus destined to be a sheep-raising region-at least during the sixteenth century (Chevalier 1975:133-135).

The resistance to or acceptance of Old World species was not only based on perceived need, ecosystem compatibility, and government order, however, it also depended on the process by which the new cultural element entered the society. Sheep-grazing was new, it did not replace a traditional system of resource exploitation; and the processes by which pastoralism developed parallel to the villageencomienda system of production meant that it did not unduly disturb the organization of traditional agriculture, even if it helped condense the space for that endeavor. It is an intriguing fact that there was really very little opposition to the initial introduction of livestock, and this despite the fact that cattle and horses, goats and sheep were all grazed within the same space used by the Otomí for agriculture and were corralled next to their houses at night. Livestock raising did not (theoretically) compete with agriculture because grazing animals exploited a previously unexploited (at least from the point of view of the Spanish pastoralists) resource, grass, and pastoralists occupied previously unknown socioeconomic niches.

When opposition to the expansion of grazing did appear, it developed in the context of access to vegetation grown in one particular place: the planted agricultural field. As long as the numbers of domestic livestock remained low, and as long as they ate only grass, there was little difficulty. But domesticated livestock did not distinguish between cultigens and other forage, and Spanish pastoralists were not terribly interested in maintaining indigenous agriculture at the expense-as they saw it-of their rapidly increasing herds, unless forced to do so by viceregal order.15 Furthermore, the fluid power relations inherent in the conquest situation meant that, wherever possible, the Spaniards ignored both Indian needs and viceregal orders and let their animals forage in planted fields.

The Otomí agriculturalists, at least at first, had no knowledge of their rights vis-à-vis the pastoralists, and could not effectively defend themselves against the invading animals. Viceregal orders, such as those of Viceroy Velasco in the 1550s, attempted to codify the timing and use of agricultural fields by grazing animals (Chevalier 1975:133-135). But it was really only with time and experience and many lawsuits-and, most importantly, with the involvement of indigenous peoples such as the Otomí in livestock raising-that the relations between agriculture and pastoralism were codified by local custom and use.

Far more threatening to Otomí agriculture than the actual adoption of pastoralism was the process of environmental transformation that accompanied the expansion of sheep into the Valle del Mezquital.

Forest clearance, burning, and overgrazing degraded the catchment area of the Valle del Mezquital and destroyed the basis of initial Otomí success in this arid region. As the water table dropped and springs failed, landowners and villagers fought each other in and out of court over the reduced water sources, drawing up agreements that usually spelled out the weakening power of villages to control this essential resource. The extent to which the catchment value had deteriorated in the Valle del Mezquital by the end of the sixteenth century is indicated by the fact that the Otomí had to fight for water to irrigate fields that were vastly reduced in extent as a result of the demographic collapse (ca. 92 percent by 1600).18 But, to the considerable chagrin of the Spaniards, the Otomí, though reduced in number and with reduced agricultural fields, managed to retain control over the best lands in the region: the humid bottomlands.

The environmental degradation that accompanied the development of intensive sheep-grazing in the Valle del Mezquital resulted in the deterioration of the carrying capacity of the range and the collapse of intensive pastoralism. Small-scale intensive sheep-grazing, with large numbers of small landowners grazing huge numbers of animals in common, developed in the 1550s and 1560s, reached its apogee in the 1570s, and began to decline rapidly thereafter; by the last two decades of the sixteenth century, much more land was needed to maintain a single head of sheep. The only way to maintain profitable flocks was to acquire several stations and to enforce exclusive access to the pasture. Those who could afford to buy land did so, and by 1600 regional production was dominated by extensively grazed latifundia. Pastoralists who had been known as señores de ganado (livestock owners on a grand scale) in the period of expansion, were now known as terrateñientes (landowners).

Another consequence of environmental transformation was a shift in regional production from sheep to cattle and goats in the seventeenth century (Melville 1994b). But the Otomí did not take part in either cattle-raising or goat-herding on the same scale as they had sheep-grazing in the era of expansion. Indeed, their marginalization can be ascribed, in part, to the depression of sheep-grazing in this region after the sixteenth century. Restrictions on the number of animals Indians could graze, and on the number their abbatoirs could butcher, also limited the ability of the Otomí to actively engage in the pastoral sector of the economy by the end of the century (Dusenberry 1963:111; Chevalier 1975:138).

In the early stages of contact with the Spanish, the Otomí had taken advantage of opportunities provided by a rapidly changing political economy in order to maintain their economic and political status in relation to the newcomers, and perhaps to leave old obligations behind. The evolution of pastoralism as a parallel system of resource exploitation allowed them to adopt livestock-raising while maintaining traditional agricultural systems. The importance of this is very clear when one considers the problems that the Spaniards faced when they tried to introduce wheat production. The Indians of New Spain persistently and successfully resisted the efforts of the Spaniards to force them to grow wheat. It was even difficult to have wheat grown as an item of tribute.

Although wheat was as crucial an element of Spanish culture as sheep-or hog-raising, and was as profitable a sector of the colonial economy, it was not readily accepted. Wheat was not an add-on in the way livestock was-it was a replacement. As a replacement it required extensive changes in many areas of social and cultural life, so that in most places it was rejected (Gibson 1964:322-323).20 But livestock could not simply be added to the culture and landscape of the Valle del Mezquital: pastoralism brought with it profound changes in the environment that, in the end, destroyed the basis of small-scale intensive pastoralism and, as well, the basis for Otomí participation in the colonial political economy, and produced a landscape within which the Otomí villages became islands in a sea of mesquite. Finally, then, one may argue that the persistence of Otomí land-management techniques after the sixteenth century resulted from cultural involution and structural-and physical-isolation.

A SOCIALLY AND CULTURALLY DIVERSE ARCTIC

"Arctic societies" and "Arctic cultures" are not as easy to define as it would appear. Seven of the eight Arctic countries have the majority of their territory and citizens south of the Arctic. If we use the widespread definition of societies as groups of human beings that have the capacity to self-reproduce their collective existence, it becomes almost impossible to speak of "Arctic societies." No human collective in the Arctic today is able reproduce itself without non-Arctic input. Culture, often defined as sets of rules and values shared by a given society, fares not much better in that respect, since all contemporary Arctic cultures are influenced by southern rules and values.

Arctic societies thus refer to groups ranging from co-residents of a settlement to ethnic groups and nations. Accordingly, Arctic cultures refer to the shared rules and values of these very diverse societies.

Finally, the notion of identity also has come to play a major role in assessing social and cultural developments in the Arctic in recent years. Identity refers to the ways in which individuals and groups perceive and act upon the social and cultural traditions they inhabit. Within the diversity of cultural and social traditions in the Arctic, one can distinguish two broad types which correspond to two waves of colonization of the Arctic: One by the indigenous peoples, who established themselves in the Arctic millennia ago; The second by peoples from a European background, whose presence in the Arctic is much more recent and who remain closely connected to societies south of them.

Not all peoples residing in the Arctic today fit into one of these two categories. There are also peoples of mixed cultural ancestries, but with distinct identities and cultures, such as the Métis of Canada and the Kamchadals and other "old-settler" groups in Russia. And there are also some relatively recent colonizers of other than European background, such as the Sakha (Yakuts).

SOCIETIES AND CULTURES: CHANGE AND PERSISTENCE

In addition, there is a third layer: recent immigrants who were born and educated outside the Arctic and who live by the cultural and social standards of their region of origin. Many return south after a few years, while some stay and integrate into the surrounding societies. In some areas of the Arctic, this recent immigration has come to constitute the majority of the population. This has been a particularly striking feature of the Russian Arctic throughout Soviet times. Even more recently, the countries of origin of these newcomers are becoming increasingly diverse. Although this chapter will address all Arctic residents, more attention will be devoted to indigenous societies and cultures than to later immigrants. There are two main reasons for this. One is that, except in Europe, the literature describing social and cultural processes in the Arctic from a local perspective has focused on indigenous communities. Although more recent settlers figure prominently in national statistics, qualitative data about their communities are rare. For groups of mixed cultural ancestry, as well as for recent immigrants, data are even less readily available. The other reason is that rapid social and cultural change has affected small-scale indigenous societies much more heavily than other groups.

RAPID CULTURAL AND SOCIAL CHANGE: TRADITIONS IN TRANSITION

Although change has been and remains a constant in the Arctic, the nature and the rapidity of the changes vary widely over space and time. Social and cultural change started accelerating around World War II, more so in the Arctic than in many other areas of the world. In its first phase, state encroachment affected the autonomy of indigenous peoples. This was followed by successful struggles to regain control over local affairs but under wholly new circumstances that included not only traditions but also modernity and globalization. This section focuses on how this encroachment has affected Arctic societies.

CHANGE: A CONSTANT IN THE ARCTIC

Human colonization of the Arctic is comparatively recent. It started at least fifteen millennia ago. Since then, marked fluctuations of the environment, some of which have been at least as rapid as those predicted for current global warming, have regularly forced human populations to adapt. In such an environment, they had to perpetually fine-tune their adaptations, or risk dying out. There are many such examples in the archaeological and historical record.

Other "internal" drivers for social and cultural change were technological innovations, such as dog traction, whale-hunting gear and knowhow, and intensive reindeer herding, as well as contacts with other populations.

A major outside source of change was the increased colonization of indigenous lands by people of a European cultural tradition. The wave of colonization started before the end of the first millennium, and despite setbacks such as the demise of the Norse in Greenland around the 15th century, it increased exponentially. In the 20th century, it included all regions of the Arctic that were inhabited by indigenous peoples. By the outbreak of World War II, nations centered outside the Arctic had established sovereignty over most of the Arctic.

INDIGENOUS?

Introduction, Arctic countries have different words to designate the peoples who were already established at the time people of European tradition came. This chapter uses the word indigenous as a common term, defined as follows:

Indigenous peoples are those peoples who were marginalized when the modern states were created and identify themselves as indigenous peoples. They are associated with specific territories to which they trace their histories. They exhibit one or more of the following characteristics:

- they speak a language that is different from that of the dominant group(s),
- they are being discriminated in the political system,
- they are being discriminated within the legal system,
- their cultures diverge from that of the remaining society,
- they often diverge from the mainstream society in their resource use by being hunters and gatherers, nomads, pastoralists, or swidden farmers,

- they consider themselves and are considered by others as different from the rest of the population.

POPULATION CONCENTRATION AND LOSS OF AUTONOMY, CA. 1940-1970

The World War II era can be considered a watershed in the history of the Arctic, initiating social and cultural change at an unprecedented rate. The point of departure for this accelerated change was neither a pristine indigenous life in isolation from the rest of the world, since such conditions did not prevail anywhere by that time, nor a long-standing period of stability or slow evolution. But for many peoples, it meant a change to totally new concepts and living habits. Before World War II, many Arctic societies led a relatively autonomous way of life based mostly on hunting, herding, fishing, and gathering, where kinship represented the main and almost sole focus of social organization.

The war itself had consequences everywhere in the Arctic. Greenland, as well as Iceland, were cut off from Denmark and had more contacts with North America, whereas the Faroe Islands were occupied by the British. Vast areas of the Saami homeland in Norway and Finland were burned by the retreating Germans. Later they were rebuilt according to new standards. The Skolt Saami left their homeland in Russia to settle in Finland Alaska saw an increase in military activity following the Japanese invasion in the Aleutians. The Canadian Arctic and the Asian part of the Russian Arctic were less directly affected, but in all regions, the war ended the Arctic's relative isolation from centers further south.

The subsequent Cold War (1948-1988) perpetuated political interest and military presence. In addition to the increased military activity, there were several aspects of encroaching modernity that drove this early phase of rapid transformation. One of them was the rush to exploit the non-renewable resources of the North, generally with little regard for environmental consequences or impacts on indigenous societies. Another major factor of change was the spread of welfare state policies. Directed at indigenous peoples, they came out as paternalistic attempts at "social engineering." In several countries, the goal was to assimilate the indigenous populations as exemplified by the policies of

Norwegianization, Russification, and Canadianization. In the Soviet Arctic, planned social change was based on Marxist-Leninist ideology. As described in more detail in Chapter 5. Political Systems, struggles for emancipation and local empowerment were already emerging in some regions before the war, beginning with the islands of the Atlantic.

At ground level, changes were numerous and pervasive. The influx of immigrants from the south accelerated. In indigenous communities, they came to fill the newly opened positions as administrators, teachers, health professionals, construction workers, etc. Most immigrants considered themselves superior to locals, at least implicitly. They enjoyed higher pay and a better standard of living. The influx of immigrants was especially strong in the Russian North, where the ratio of natives to incomers was transposed, going from 10:1 to 1:10 in some regions. In Greenland, the Danish population increased nine-fold between 1950 and 1975, when it amounted to almost 20% of the total population of the island.

In all parts of the Arctic, the population became less scattered. This centralization involved relocations of whole groups of peoples, some of which were imposed. Previously nomadic peoples became sedentary, and "unprofitable" settlements were closed down. With a few exceptions (e.g. the Nenets reindeer herders of the Yamal Peninsula in Russia), centralization of nomadic populations was achieved by 1970. In the centralization process, new communities were built where many functions were controlled by incomers. A trend of urbanization was also emerging with people concentrating in a smaller number of larger settlements and towns.

In the Canadian Arctic, it remains a matter of contention whether the process of concentration was imposed or corresponded to the wishes of those involved. In many regions it entailed both aspects to some degree, often in succession. Although many individuals felt helpless and behaved passively in the midst of these transformations in and around their lives, local leadership developed, and these new leaders eventually took over responsibilities in the community.

Partly as a consequence of the availability and the form of new housing, household size and composition tended to shift from the extended to the nuclear family. In Greenland for instance, the average household size decreased from 7.6 in 1901 to 5.3 in 1955 and 2.6 in 2003. In some communities, there has been a development towards "matrifocal" families, revolving around a woman, rather than a couple, as head of household. Around 1975, 20% of all Asian Yuit (Russian Eskimo) families were headed by women, while about a third of the males aged 25-45 remained single.

Mandatory school education was introduced among some indigenous peoples in the 1950s or 1960s. For some children it meant attending boarding schools away from their families, where they were inculcated with foreign languages and cultures. In the process, many lost fluency in their mother tongue and were alienated from their families and communities.

Health care services were vastly improved with life in settled communities. As a result, mortality decreased while life expectancy and fertility increased, causing rapid population growth. At the same time, traditional healers, 48 Arctic Human Development Report healing methods, and midwifery were marginalized. In some regions, wage employment and a cash economy were introduced when people settled into modern communities. This meant that kinship ties ceased to be the exclusive focus for economic cooperation. Moreover, unemployment also became a feature of life in settled communities, exacerbated by the rapid population growth. As it became increasingly possible to make a livelihood from activities other than traditional ones, the share of traditional activities in the local economies decreased. The symbolic value of hunting, fishing, and herding has been maintained or even increased , however which is not surprising given that work and subsistence activities are a central aspect of identity.

Transport and communications improved at an increased pace from the time of World War II. Television has become available in most Arctic households in recent decades, with its attendant exposure to new lifestyles and role models.

Most of the social and cultural changes in the post-war era were directed by government agencies. Many of them, such as the provision of education, health care, and welfare, had ethical justifications. Even though some of these changes were for the better, the way they were imposed led to a loss of control over local affairs, and over collective as well as individual destinies. In small northern settlements, local people could not escape the impression that they were watching helplessly while things were being done around them and "for" them. As former Greenland Home Rule Premier Jonathan Motzfeldt has put it: "... things were administered by Danes, decisions were taken by Danes, and problems were solved by Danes. [...] The common Greenlander had a feeling of standing outside, of being observer of an enormous development, which s/he did not have the necessary background to understand".

Even a seemingly more benign intervention, such as the imposition of a Euro-American system for family names among the Canadian Inuit during the 1960s and 1970s, introduced important departures in representations of selves and social relations. The feelings of estrangement caused by the loss of control over changes in turn contributed to the rise of social problems, such as suicide, violence, law-breaking, and alcohol abuse.

The decades around the middle of the 20th century were for many northern indigenous peoples the period in their histories when they had the least autonomy, and, simultaneously, were undergoing the farthest-reaching changes they had ever experienced. Some have evoked the notion of a "lost" or "broken generation" to describe the group of people who lived "between two worlds", not really belonging to either of the two, and who may have felt as little connected to their parents and grandparents as to their own children.

Alexina Kublu tells of her experience of growing up in Igloolik and in Igluligaarjuk (Chesterfield Inlet), where she attended residential school from 1962 to 1967. She is currently Senior Justice of the Peace in Nunavut and chair of the Akitsiraq Law School Society.

The first part of my childhood was spent close to my parents' families. As my grandparents didn't live at the same place we were surrounded either by paternal aunts and uncles or maternal aunts and uncles and their cousins as well as all my cousins wherever we happened to be. As children there was a "blanket" treatment of us yet we were treated as individuals. There were aunts and uncles to dole out affection, others to discipline, others who were childhood friends and others who ordered you around to help out with chores and you felt proud to be considered able to help.

The latter part of my childhood was spent in the residential school in Igluligaarjuk. You all got to do chores as they were assigned and not because someone asked that you do it. You did not feel the pride of accomplishment but felt the wrath of failure to do so. The life you lived at home and the life at the school were so alien to one another you lived them separately and did not incorporate the lifestyle of one with the lifestyle of the other.

From 1960 to 1987, the Inuit population of Igloolik grew from 87 to 806, the White from 4 to 48, tenfold increases. The percentage of extended households decreased from 76% to 10%, to the profit of nuclear households. Many jobs were created. Simultaneously, the importance of hunting decreased. Alcohol and drug consumption increased, as well as criminality and cases of suicides.

Igloolik (2003 population: 1457) is home to Isuma Productions, which produced Zacharias Kunuk's internationally acclaimed movie "Atanarjuat, The Fast Runner" (2000). Growing up with change in the Canadian Arctic Regaining autonomy and reaffirmingcultures and identities In all of the Arctic countries whose governments were willing to negotiate land claims and measures of autonomy, indigenous peoples were quick to present coherent demands that reflected their cultures and aspirations. This demonstrates that indigenous cultures and identities had not been entirely crushed. Even during the "dark years" of directed change, an elite was able to emerge which was conversant with, if not at ease in, "both worlds." These developments took place within the framework of nation states and in contexts of increasing economic dependency and encroaching globalization.

With the hope of regaining some autonomy came the reaffirmation of identities, cultures, and sometimes also languages. As Nikolai Vakhtin puts it about Soviet and post-Soviet Chukotka, "the horrible totalitarian mincing machine in which the Chukotkan minorities found themselves in the 1950s-1980s could decrease their ability to resist, could decrease the linguistic and cultural differences between them, but was far from what is needed to wipe those differences (and those people) out. As soon as the leveling pressure went down, immediately new groups, new types of individual and group identities, even new languages, began to appear." Although it does make reference to roots in the past, cultural reaffirmation is not a "return to traditions" in the sense of a simple reactivation of previously existing customs. It is an active re-creation of culture and symbols, whose functions in current contexts differ from the ones they had a few decades earlier.

The context for this cultural reaffirmation is one in which most indigenous peoples have been living since at least the 1970s in settled communities, surrounded by most of the paraphernalia of modern life in the way of household goods, communication technology, modes of transportation, but also community committees, local schools, recreation, etc. Many of the challenges they face are therefore new ones. Social unity has been harder to achieve in communities that bring together groups that had little to do with each other in earlier times.

Societies have also become more complex and more differentiated, and are segmented along new lines, such as lifestyle patterns, political parties, and religious denominations. Social stratification has increased. Sharing networks are affected. And, as much as anywhere else, gender roles are in the process of redefinition.

As we discuss later in the chapter, some indigenous languages have disappeared, but others are being revived, and others still are as healthy as ever. Group identities have shifted, but their distinctness has been preserved.

World views and spiritualities are reaffirmed. The arts are flourishing, often as vehicles as much for collective as for personal identity. More and more indigenous persons live in urban centers outside the Arctic. More often than not, they expand the borders of their home communities rather than getting "lost in the crowd." Less trivial than it seems, food tastes have changed, which has an impact on the degree of local renewable resources exploitation, on the dependence on imported foods, and on household activities and budgets.

HOW MUCH CHANGE CAN CULTURES AND SOCIETIES TAKE?

Change is an aspect of all societies and cultures. It is more "normal" and in that sense "healthier" than stagnation. Traditions come, transform themselves, and some eventually pass. Obsolete elements of cultures are discarded and new ones emerge. One must question the tendency to consider change as a threat to some immemorial "tradition" in discussing indigenous societies, when it is called progress in western societies.

How the people affected by changes perceive them (the so-called emic perspective) determines how they react to them. Research so far has focused too little on local perceptions of changes. One may strongly suspect that the most damaging aspect of change is the feeling of not being able to control it locally, neither individually, nor collectively as a society. This means that attempts at "planned social change" from the outside are doomed to create as many problems as they solve, even if the planning rests on pertinent scientific knowledge.

The occurrence of massive changes in all aspects of life simultaneously with symptoms of cultural and social distress (e.g. violence, suicide, alcoholism) has made many people suspect a causal association between the two: the symptoms are seen as the consequence of the inability to cope with the changes. Little research has been done so far, however, into what may be considered "overly" rapid transitions and the accumulated effects of different types of changes. Therefore, we still lack convincing explanations of why problems erupt when and where they do, such as in the traditional and isolated

communities rather than in larger centers where the pace of change is more rapid. We are even less able to predict the outcomes of change. For example, local empowerment does not in and of itself reduce vulnerabilities. This is illustrated not only by the fact that suicide rates are comparably high in Nunavut and in Greenland, although these two regions have followed quite different historical developments but also by the fact that rates have continued to increase ever after the introduction of self-government.

Health specialists tend to explain the symptoms in terms of response to "acculturative stress", i.e. a view of indigenous cultures and identities as bound to erode into those of the mainstream. This does not fit the fact that Arctic indigenous cultures and societies have proven to be resistant and resilient.

TOWARDS A COMMON ARCTIC CULTURE AND IDENTITY?

Recently, the spread of information and communication technology has given rise to new blends of traditions and elements of "world culture," in music and arts, clothing fashions, food, etc. This coexistence of traditions and modernity is currently observed among many indigenous peoples worldwide. It includes renewal and in some cases reinforcement of ethnic identities, as well as an instrumentalization and commoditization of cultures. This trend is probably stronger in the Arctic than in poorer regions of the world because of the generally high incomes and good communications networks. These new blends illustrate Marshall Sahlins's assertions, based on recent research in many parts of the Third and Fourth Worlds, that peoples around the world see no opposition between tradition and change, indigenous culture and modernity, townsmen and tribesmen.

Culture is not disappearing, he concludes: rather, it is modernity that becomes indigenized. In the North, this process includes elements of an emerging Arctic identity. The Arctic is generally considered as a specific geographic and cultural area and within each of the countries that straddle the boundaries of the Arctic, "nordicity" emerges as an element of regional identity. In some cases, indigenous peoples may downplay their local identity and emphasize a common indigenous background. For example, the Inuit Circumpolar Conference and the Saami Council have fostered, respectively, common Inuit and Saami identities. In some regions, indigenous and non-indigenous identities are converging. Increased cooperation across the Arctic may foster the emergence of a feeling of community throughout the region.

CIRCUMPOLAR SYNCHRONICITY DESPITESHORT-TERM OFFSETS

Within the overall trend of rapid social and cultural change, there is variation across the Arctic as to when these changes occurred. In Fennoscandia, indigenous peoples' contact with people of European origin started a millennium ago, whereas this happened only a century ago in remote areas of the Canadian Arctic. Developments in the Soviet and post-Soviet Russian Arctic have been quite distinct. Widescale changes started already a decade before World War II, with massive efforts to exploit nonrenewable resources, importing immigrant populations for that purpose. In addition, the state implemented ideologically driven experiments aimed at transforming society, especially with regard to indigenous peoples. It was not until 1990 that Russian indigenous peoples were allowed to create their own independent national organization: the Russian Association of Indigenous Peoples of the North (RAIPON).

Small indigenous peoples of the Russian North are now better connected with developments abroad than they were in Soviet times, but the evolution of their situation still remains quite distinct from other circumpolar regions. Except in resource rich regions, this situation is characterized by dis-investment from the state, which previously played a central economic role. This has caused severe impoverishment and social and cultural disruptions. In addition, there has been a massive out-migration of recent immigrants. Ever since the collapse of the Soviet Union, the Russian state has cut subsidies to its Arctic provinces, which has been very effective in inducing newcomers to leave as in Chukotka. Without such economic (dis-)incentives they tend to stay: Norilsk is a case in point.

Another significant area of contrast is that whereas most parts of the Arctic were Christianized before World War II, this process only began a decade ago in Northeastern Russia. During the 20th century, differential access to self-government introduced further divergences in social and cultural change across the Arctic.

TREND SUMMARY

In spite of some differences in content and the timing of changes across the Arctic, some trends of the past few decades are amazingly similar and synchronous. Arctic cultures and societies are rapidly being transformed. After World War II, the rate of change accelerated and paternalistic welfare state policies were imposed. Such planned change gave rise to hitherto unknown social problems. Later, as the pressure to assimilate decreased, cultures and identities started to reaffirm themselves. This revival of tradition occurs in new forms that accommodate modernity and incipient globalization.

CULTURAL TRENDS: LANGUAGES, RELIGIONS, WORLD VIEWS, AND ART

The rapid social changes that have taken place since World War II, coming on top of the fact that many Arctic societies remained isolated from other parts of the world for so long, have had profound effects on cultural expressions such as languages, religions, world views, and art. This section provides a survey and some examples of the current situation across the Arctic. A recurring theme is again the merging of traditions and new influences.

LANGUAGES: LOSSES AND REVERSED LANGUAGE SHIFTS

The circumpolar North today is dominated by languages that originated far south of the Arctic. Russian is currently the most wide-spread language, followed by English, Norwegian, Icelandic, Swedish, Finnish, and other Scandinavian languages. The gradual spread of these languages, brought to pass by the Euro American colonial expansion of the past millennium, peaked in the 20th century as a result of mandatory schooling for all Arctic residents.

Many of the over 40 indigenous languages, which had characterized the linguistic space of the Arctic for centuries or even millennia, became threatened by extinction or insignificance after World War II. Icelandic and Faroese, which despite their southern roots can be said to have developed into their present form in situ, are exceptions. Apart from their ambiguous status as being indigenous to the Arctic, however, these languages are also characterized by a retention rate of almost 100%.

The state of indigenous languages throughout the circumpolar North varies significantly. Many are characterized by a dramatic loss of speakers, including many of the Saami dialects/languages, Yukagir, Aleut, and several Athabaskan languages in Alaska and Canada. Other indigenous languages seem to be in relatively good shape. These include Northern Saami, Tundra Nenets, Sakha, Chukchi, St. Lawrence Island Yupik, central and eastern Canadian and Greenlandic Inuktitut (the eastern variants of the Inuit language), Chipewyan, Dogrib, and Slavey. One obvious reason for this difference relates to demographic factors. It is not surprising, for example, that the language of the Sakha, who number over 400,000 (although only a small portion of them live in the Arctic), is in much better shape than the language of the Yukaghir of whom there are only slightly more than 1,000 people. A variety of other factors, including government policies, ethnic prestige, and local leadership, have also played a role in language retention.

Government policies throughout the circumpolar North were generally indifferent, sometimes even hostile to the preservation of indigenous languages well into the second half of the 20th century. The situation in the Soviet Union was different and asynchronous with other parts of the Arctic. On the one hand, indigenous languages of the North were actively supported back in the 1920s. On the other hand, local initiatives to battle language loss had little chance under the authoritarian conditions of pre-perestroika politics.

Danish Government policies regarding the Greenlandic language though long based on paternalistic attitudes represent probably the longest history of language preservation, with the result that the Inuit language-particularly as spoken in Greenland-is one of the more positive examples of Arctic language retention today. The box below describes this case in more detail.

Greenlandic is part of the Inuit language and of the Eskaleut linguistic stock. The Inuit language is spread from the eastern tip of Siberia and Aleutian Archipelago through the whole Arctic coastline of North America, the Labrador Peninsula, and Greenland. The basic grammatical structures, the phonological and cultural base are virtually the same. Regionally different external influences through European contact led to distinct educational policies for each area. The Greenlandic experience was "softer" than those of Inuit in other areas. Not only were the Dano-Norwegian policies of the 18th century "nicer," but missionaries tried to reach the Inuit in their own language, and trade relations with outside merchants became more influential.

Certain intellectual trends also influenced the policies of Denmark towards the Greenlanders. For example, between the 1840s and the 1860s, progressive administrators and young intellectuals facilitated the development of Greenlandic literacy and educational materials. S. Kleinschmidt developed a standard system of spelling for the Greenlandic Inuit language, issued a Greenlandic dictionary, and provided a full translation of the

Bible, while H. Rink published traditional Greenlandic stories and tales. At the same time, the first Greenlandic periodical was published and the principle of political self-government was gradually introduced.

A new era started in the 1950s. Greenland was incorporated into the Danish jurisdictional sphere on Danish terms and Greenlandic was challenged by the massive influx of the Danish language and Danish educational practices. There were cases when Greenlandic parents encouraged their children to speak Danish even though they did not understand Danish themselves. This period lasted until Greenland Home Rule was introduced in 1979. The new political situation fostered the stabilization of the Greenlandic Inuit language. The Home Rule Government was in charge of education in Greenland and the Greenland Home Rule Charter made Greenlandic the main language, while providing for Danish language teaching and permitting the official use of both languages. In the meantime, a new phonemic orthography (a writing system which tries to achieve a close resemblance between spoken sound units and written symbols) had been introduced in 1973, which made it much easier to write in Greenlandic.

Poetry and music are now blossoming and literature is being revitalized. The Greenland Home Rule Government established a Greenland Language Committee, a Greenland Place Names Authority, and a Greenland Committee on Personal Names. Those committees are responsible for tracking language and naming trends and also collaborate with similar authorities in neighboring and Arctic areas.

Carl-Christian Olsen (Puju), Greenland Home Rule Language Commission Hopes and motivations for the Greenlandic Inuit language Contact between indigenous and outside languages in the Arctic has been going on for centuries. One result has been the development of a number of pidgin and contact languages, as well as of one "mixed language," namely Copper Island Aleut. During the 20th century, however, language contact reached a new level of intensity, leading to one-sided language shifts in favor of southern languages promoted by schools and government agencies. Some people became bilingual, but it should be pointed out that most Arctic communities had been bi-or even multilingual long before southern states encroached onto their territories.

Within the general picture of language loss, there are fortunately also examples of reversed language shift, where indigenous languages are regaining ground. This includes the Saami language, where the case study presented in the box below illustrates how political, social, and cultural empowerment enables language revitalization. A similar development occurred in the 19th century when political and romantic nationalism stimulated a revival of standardized and purified Icelandic and Faroese. These examples confirm an earlier observation that language shift is not about languages but are a social, political and economic phenomenon.

An important question is how individual speech communities deal with language shift and loss. Does the apparent link of language and identity lead to the disappearance of ethnic and other group identities as a result of language shift processes? Luckily, there are communities and groups, such as the Inuktitut speakers of Nunavik, where almost 100% of the population continues to speak the language and where it fulfills a variety of crucial functions in contemporary life. On the other hand, for many Alaskan Iñupiat, indigenous language use is limited to individual words and phrases and a particular intonation of Village English. Still, even in the Iñupiaq case, "language may continue to fulfill its [symbolic] functions even in its absence". Thus, while many Arctic indigenous languages may lose some of their communicative functions to English, Russian, or other wide-spread languages, individuals and groups will continue to use language as symbolic markers of identity.

RELIGIONS AND WORLD VIEWS: MERGER OF TRADITIONS AND CHRISTIANITY

At the beginning of the 21st century, the vast majority of all Arctic residents are affiliated with some form of Christianity. Various Protestant churches dominate in northern Fennoscandia, Iceland, the Faroe Islands, Greenland, Alaska, and parts of northern Canada, while the Russian Orthodox Church is prevalent in the Arctic regions of the Russian Federation and in addition has limited presence in Alaska and parts of Finland. Finally, the Roman Catholic Church is particularly strong in parts of Canada and Alaska. There is considerable variation as to when Christianity reached different parts of the Arctic. While it happened almost 1,000 years ago in northernmost Europe, the inhabitants of the Chukchi Peninsula in the Russian Far East had little first-hand experience with Christianity before the 1990s. Generally speaking, the 18th and 19th centuries were the major periods of religious conversion in the Arctic.

There have been various attempts to explain the rapid conversion of Arctic peoples to Christianity. Most authors agree that a combination of several factors is responsible. In some cases, such as the Lule Saami, religious change was forced, but in many other cases indigenous agency was much more pronounced.

The adoption of Christianity rarely, if ever, resulted in the simple replacement of one religious system by another. Instead, old and new beliefs were reintegrated within a new system that was both Christian and local. In Russian Orthodoxy has been a prominent part of Aleut life since approximately the 1790s, when the first missionaries arrived and established churches in most Aleut communities.

While some researches have highlighted the devastating effects of the first fifty years of Russian-Aleut contact (1741-1790s) and the role that the church played in imposing new social forms, religious institutions, and material culture, others have pointed to the similarities between the Aleut pre-contact religion and Orthodoxy as the reason for the acceptance of the new religion.

The Russian Orthodox Church triggered undeniable accomplishments in education and literacy by supporting elements of traditional Aleut culture and by preserving their language. This enabled a peaceful intertwining of both cultures. For the majority of contemporary American Aleuts, the Russian Orthodox Church is woven into the fabric of their lives. They are devoted followers and believe it was embraced by their ancestors over two hundred years ago. Many communities carefully maintain graceful church buildings that are not only places of worship but also symbols of Aleut culture.

The Commander Island Aleuts traveled a different historical path. Their islands were resettled from the islands of Atka and Attu in the western Aleutians in the 1820s. These islands were not included in Russia's sale of Alaska to the Unites States in 1867, and the people were separated from their original communities. After the Russian Revolution of 1917, they were converted to atheism by the Bolsheviks. When the Orthodox Church began regaining influence in the changing post-Soviet society in 1990s, the Aleuts of Russia came full circle. The Russian Orthodox Church is not the only religion on the Commander Islands today and the impact of religion on the social lives of the Aleuts is still limited. An Orthodox parish was formed in the village of Nikolskoye in 2000 and an apartment converted into a "church." Occasionally, priests from nearby Petropavlovsk come to perform weddings, baptisms, and other functions.

An Aleut Elder, Vera Timoshenko, recollects how her mother told stories about the closely knit community of Nikolskoye and the role of the Orthodox Church in teaching people how to be humane and compassionate, industrious, and obeying. Vera wishes Nikolskoye would build a real church to revive the Orthodoxy that used to be a spiritual foundation of the Commander Island residents. She believes that the rebirth of the church may solve some of the social problems, such as alcoholism, and may teach people to treat each other with kindness and respect.

Victoria Gofman, Executive Director, Aleut International Association Aleuts and the Russian Orthodox Church some cases, a religious tradition that was initially introduced through colonial expansion, not only became part of, but even reinforced the cultural identity of an indigenous people. One example of this can be found in the Aleutian Islands, as illustrated in the box on page 56. Pre-Christian religious beliefs in the Arctic showed an abundance of local and regional variation. Nevertheless, two broad categories of belief systems-shamanism and animism-were characteristic of most of them.

Shamanism, which is often perceived as the stereotypical "Arctic religion" by outsiders, was never a unified system of beliefs but a variety of localized practices with a limited number of common elements. Central was the figure of the shaman, sometimes male, sometimes female. Until recently, most Arctic communities had religious functionaries who were able to communicate with and "control" spirits. These shamans were engaged in healing and other activities aimed at improving communal and individual well-being. In the small-scale societies, these functionaries held extremely important social positions, which sometimes led to an abuse of power.

Animism is the belief that all natural phenomena, including human beings, animals, and plants, but also rocks, lakes, mountains, weather, and so on, share the soul or spirit that energizes them. This notion is at the core of most Arctic belief systems, which means that humans are not the only ones capable of independent action. An innocuous-looking pond, for example, is just as capable of rising up to kill an unsuspecting person as is a human enemy. Another fundamental principle of Arctic religious life is the concept of humans being endowed with multiple souls. The notion that at least one soul must be "free" to leave the human body is basic to the shaman's ability to communicate with the spirits.

Since the killing and consumption of animals provide the basic sustenance of circumpolar communities, ritual care-taking of animal souls is of utmost importance. Throughout the North, rituals in which animal souls are "returned" to their spirit masters are widespread, thus ensuring the spiritual cycle of life. While most animals of prey receive some form of ritual attention, there is significant variation in the elaboration of these ceremonies. One animal particularly revered throughout the North is the bear.

Religious beliefs and practices in the Arctic have always been tied to the land and other aspects of the visible and invisible landscape and "sacred sites" are of particular importance for many Arctic residents.

Sacred sites are frequently located in regions where nature preservation has a high importance for the indigenous peoples: on highly efficient hunting grounds, in regions with rich biodiversity, along migration routes, in areas populated with rare species, as well as in areas with unique landscapes. The Program on Conservation of Arctic Flora and Fauna (CAFF) recently supported a proposal by the Russian Association of Indigenous Peoples of the North (RAIPON) to implement a pilot project in

Russia to protect indigenous peoples' sacred and ritual sites. The aim is to promote the preservation of indigenous cultural and spiritual heritage in the Arctic in sites that have sacred significance and enjoy high biodiversity and nature value, in order to demonstrate the necessity to incorporate sacred sites into the network of protected natural territories.

In addition to literature and archive reviews on topics such as indigenous culture, spiritual and religious representations, traditional knowledge, resource use, and environmental protection, experts from CAFF and the Indigenous Peoples' Secretariat have developed a questionnaire that regional researchers and local, indigenous assistants have used to identify sacred sites. The results of the field studies were presented after discussions with the communities involved and the idea is to develop detailed recommendations to governmental authorities in Yamal and Kamchatka. In the Tazovsky district of the Yamal-Nenets Autonomous Okrug, 263 sacred sites were identified, described, and mapped from interviews with indigenous elders.

In the Koryak Autonomous Okrug, interviews in villages of the Oloyutorsky district helped describe and map 84 sacred sites. All questionnaires are archived in the RAIPON office and can be used for further research. The project will also identify ways to improve the protection of sacred sites. In conclusion, it seems that modern civilization is finally learning from indigenous peoples, who still worship Nature as a living being and consider contacts with the land to be the primary factor of survival. One of the project participants expressed it as follows: "Sacred sites mean environment, morality and veneration for life."

The Conservation Value of Sacred Sites: a Case Study from Northern Russia

World views are not limited to the religious sphere. In fact, Arctic indigenous worldviews are characterized by their holistic nature, which means that they cannot be easily compartmentalized into religious, economic, social, or other components. Still, those aspects of Arctic world views such as "ecological" practices and beliefs that seemed non-religious to outside observers, faced less opposition from missionaries and state agencies than those more obviously directed toward supernatural forces. While it would be naïve to assume that ecological beliefs and practices have not changed over time, the persistence of "subsistence ideologies" is noteworthy. Contemporary Gwich'in thinking on "subsistence" illustrates how changing economic and religious conditions become integrated, as is described in detail in the box below.

ART: FUNCTIONAL BEAUTY AND COMMERCIAL COMMODITY

Art in the Euro-american understanding of the term-objects made solely for aesthetic purposes-did not exist in the Arctic until recently. However, an archaeological record with wonderful sculptures and drawings shows that peoples in the Arctic have been making objects that were functional and aesthetically pleasing from time immemorial.

The first encounters with outsiders provided new possibilities for artistic expression, for example by the introduction of iron tools. However, Christian missionaries and government officials were often responsible for undermining the religious basis on which most of 58 Arctic Human Development Report The Nenets sacred site Numt' khantorma.

Subsistence, narrowly defined, means to survive. To the Gwich'in, it means far more. Besides our spiritual relationship with God, the Creator of all that is, subsistence is the essence of the Gwich'in Nation. It is how we are sustained physically. It serves to support us economically and spiritually and is a key to our sustainability as a people. We are fed by plants and animals of the water, air, and land. Wood provides warmth and housing and the raw materials for tools and transportation devices, such as boats, snowshoes, and toboggans [runnerless sleds widely used by Native Americans].

The cash component of our economy is a small but important element of our subsistence lifestyle. We harvest wood, fur, and wildlife for barter or sale, mostly locally, but occasionally abroad. While some have assumed the role of full-time workers, many rely mostly on natural resources they catch or cut to meet their economic needs. Our spirituality is tied inextricably to the water, land, air, and resources within them. Our relationship to the spiritual realm has always been conducted through the beauty and awe of nature around us. We are sustained as individuals and as a people through our subsistence lifestyle. Our resourcedependent culture is sustained by unfettered mobility and access to the resources we depend on. Unfettered mobility means that we pursue resources where they are most abundant and where there is easy access. Unfettered access means that we gather resources when we need them and in the most economical and feasible manner possible. As we continue this lifestyle, we build and maintain our close and interdependent ties to the resources around us.

Gwich'in identity is a picture of integration with the land and resources. We see ourselves as an integral part of the diversity of the landscape. We believe that we would not be whole if we were separated from this land. We also believe that this land would not be whole without our presence. Our well-being is linked closely with our ability to live on and adapt with the land. Our family and land-based bonds are strengthened, restored, and invigorated as we continue our subsistence lifestyle. A tremendous sense of belonging and purpose is experienced as we survive on the land. Subsistence encompasses all areas of Gwich'in life from the cradle to the grave. Gwich'in youth are trained early on the intricacies of our relationship with the natural environment and the harvesting, processing, and distribution of wild resources. These relationships are strengthened when our youth mature into leaders. The cycle of life continues as they pass what they have learned from their elders and through trial and error to their children and relatives.

Arctic artistic production was based. In certain areas, such as Greenland, art came under direct influence of European traditions early on. The Greenlander Aron of Kangeq (1822-1869) became known throughout Greenland and Denmark for his lively watercolors of Inuit village life and tales. In other areas, such as Alaska and many parts of Arctic Canada, handicraft items for trade provided a venue for "native art." For example, delicate Athabascan beadwork on moose and caribou skin was popular throughout the Canadian and Alaskan Arctic and subarctic.

The entry of Arctic art into international markets is recent. One of the best-known examples is Canadian Inuit printmaking. In 1948, James Houston, a young non-Inuit Canadian artist, traveled north to the Nunavik village of Inukjuak for a sketching trip. Houston befriended the local Inuit, who coveted imported commodities. In trade, the Inuit brought him small soapstone models of animals.

Houston persuaded the Canadian Government to subsidize soapstone carving, which eventually became a multi-million-dollar enterprise for the Inuit. A decade later, Houston had moved to Cape Dorset on Baffin Island and repeated the same success story with printmaking. There, local Inuit artists submitted drawings for print making. The prints were marketed in North America and Europe, and the demand soon outstripped the supply. Thanks to worldwide media coverage, artists such as Kenojuak Ashevak and Pudlo Pudlat became famous with Inuit art collectors. Their works are in museums, art galleries, and private collections around the globe.

In the early 21st century, indigenous art in the circumpolar North is thriving. Cruise-ship passengers and other tourists are eager to bring home objects which signify the exotic Arctic. In Alaska and coastal British Columbia, gift shops routinely sell copies of native art mass produced in Asia where labor is cheap. The authenticity of indigenous art is to some extent protected by subsidized programs that provide artists with a sticker guaranteeing the authenticity of their work.

However, more and more artists in the Arctic do not want to be seen as representatives of a particular ethnic tradition but as active participants in a globalized art scene. Whatever the position of the individual artist is, the fact remains that almost all indigenous art from the Arctic is today created for consumption in a culture that is economically and politically more powerful than theirs.

In recent years, the development of Arctic arts has gone far beyond the confines of what have been traditionally considered the fine arts. New art forms, such as literature and filmmaking, have become prominent. For example, the critically acclaimed film Atanarjuat ("The Fast Runner")-written by Paul Apak Angilirq and directed by Zacharias Kunuk-is the first feature film made in Inuktitut. Moreover, writers such as the Chukchi novelist Yuriy Rytkheu have successfully transformed oral traditions into books which are read throughout the Arctic and non-Arctic world. Finally, new forms of Arctic music are developing, which incorporate traditional elements, such as the Sami yoik, and elements of western popular music.

SIMILARITIES AND DIFFERENCES WITH NON-ARCTIC AREAS

Many of the cultural trends in the Arctic are the result of an unbalanced encounter between the cultural traditions of small-scale, hunter-gatherer societies and large-scale agricultural and industrial states. What is peculiar for the Arctic is that these encounters occurred relatively late, and that agricultural/industrial cultural values were imposed in the 20th century. The similarities to non-Arctic areas are greatest with those of other hunter-gatherers pushed aside by agriculturalists relatively recently, as in Australia and Amazonia. However, the indigenous groups in the Arctic are generally less impoverished than in their thirdworld counterparts. And even more important, they are part of larger societies that have come to support-by and large-a fuller implementation of civil and indigenous rights.

VARIATIONS WITHIN THE ARCTIC

Various parts of the Arctic came into intense contact with cultural agents from the outside at different points in time, which in turn often determines the extent to which non-Arctic elements have been incorporated into local cultural traditions. An example is the almost complete erasure of shamanistic elements from Saami worldviews as a result of almost 1,000 years of Christian influence. For current cultural processes in the North, government policies are among the most important variables. In the 20th century, the policies implemented by the Soviet Union differed most from other Arctic countries.

Moreover, the cultural trajectories of Iceland and the Faroe Islands are noticeably different from the rest of the Arctic, primarily because of their different settlement history. While the cultural background of the ancestors of the contemporary Icelanders and Faroese was undoubtedly non-Arctic and agricultural, their descendants can point to over 1,000 years of cultural development in the Arctic.

TREND SUMMARY

Outsiders and Arctic residents have been bemoaning "culture loss" for decades. This kind of judgment fits with the measurable decline in linguistic and religious knowledge, the fact that certain songs, dances and other art forms were pushed out of use, that languages became extinct, and worldviews replaced. However, "culture gain" and "culture creation" are also part of the cultural realities of the Arctic. Vocabularies, dialects, and languages were replaced by others, as were religions and art forms. Also, many aspects of Arctic worldviews have persisted despite processes of change and replacement. In the final analysis, the most important factor is whether the local community in question identifies with the cultural bricolage its residents hold today. Culture is intimately tied to identity and the major question is whether you can consider the language/s you speak and the spiritual entities you respect as "yours," no matter where they "originated."

SOCIALIZATION, KINSHIP, AND NEW NETWORKS

The accelerated change of recent decades has been accompanied by gaps in socialization and knowledge transmission between generations. Despite these difficulties, kinship and family relations, especially for indigenous peoples, have remained a central focus. This is true also in the growing Arctic towns and even in urban settings outside the Arctic homeland. Contemporary social networks also include connections between recent Arctic emigrants to more southerly cities and the Arctic communities they came from, as well as between southern immigrants to the Arctic and their social milieu of origin. This section highlights how current social change is shaped by relations between generations, kinship networks, urbanization, and the extension of local communities through emigration.

SOCIAL TRANSMISSION OF KNOWLEDGE ACROSS GENERATIONS

Among indigenous peoples, recent changes have been so precipitous that they have been interpreted by outsiders as a break from the past, and in some cases even, as a breakdown of societies and cultures. In particular, transformations have been associated with major communication gaps between generations. These gaps were amplified by just as abrupt linguistic shifts, which are described in the previous section of this chapter. In many regions, children were in boarding schools where the only language they spoke was a language different from that of their people. Communication with their grandparents was thus often precluded. Arctic developments confirm the fact that minority languages can be dealt deadly blows in the space of one or two generations.

The break in socialization was accentuated when social changes and the school system disrupted important features of traditional education systems. For example, among many indigenous peoples, a great deal of learning occurs through observation and imitation rather than through the written or spoken word. Transmission of knowledge also takes place through the telling of myths and stories that implicitly teach lessons of life to those who are culturally trained to decode them. Furthermore, among Inuit for example, children are considered as autonomous beings, embodying the name soul of a deceased relative, whose own volition should not be interfered with.

These aspects of traditional culture have clashed with the recently introduced formal education and new role models. The introduction of culturally sensitive curricula should reconcile public education with other modes of socialization. In some cases, parents thought they were improving their children's chances by letting them be educated in another culture and language, rather than encumbering them with a cultural heritage they were told was obsolete. Such a break in communication between generations, which was common in the second half of the 20th century and continues today in some regions, has created a bottleneck for social reproduction. Nowadays, parents struggle with the atomization of the household and the competition of television programs when they try to provide role models, while indigenous elders strive to reach out to youngsters. The box below illustrates this modern dilemma.

KINSHIP AND OTHER SOCIAL NETWORKS

In the Arctic, kinship has for centuries determined the choice of marriage partners, where you live, and participation in subsistence and ritual activities. The definition of kinship and relatedness, however, does not give the same emphasis to biological relatedness as exists in most Euro-american societies. Notions of kinship are flexible and in many cases, social relations are kinship relations. Kinship structures relations of cooperation and mutual aid, as well as those of avoidance and hierarchy.

The majority of indigenous Arctic societies (e.g., most Inuit and Yup'ik societies, the Saami, and the Chukchi) are characterized by bilateral principles of descent reckoning. This recognition of both the father's and the mother's kin provides ample means for including a wide range of people among those considered relatives. Patrilineal descent is found only among the Samoyedic, Tungusic and Turkic groups of northern and central Siberia, probably an influence from Central Asia.

Matrilineal descent, on the other hand, is only prevalent in the northwestern part of the North American Arctic, among Athabaskan groups of Alaska and, to a lesser degree, Canada. Since matrilineal systems need fairly large groups of people to function properly, bilaterality among most Canadian Athabaskan groups could be due to resource pressures and population losses triggered by Euro-american expansion. Athabaskan kinship terminologies sometimes refer to particular cousins as spouses, thereby implying cultural

What child-rearing practices we follow are not that of our parents, as we were too young to have learnt it before we went off to boarding school. We did not learn to be parents from school either, because there was no parenting role model to follow. We have children who have grown up in the TV age.

They are learning what families are "supposed to be like" from what they see: mostly TV sitcoms. In addition some of our children are affected by the abuse of alcohol and other substances. We are the only ones in our self-contained nuclear household trying to be mother or father who is supposed to provide affection, discipline, enjoyment time as well as getting them to help around the house [. . .]. These are the changes that have occurred in one generation. Who is raising the children? What parenting skills have they been given? What parenting skills are they passing along to their children?

Alexina Kublu, Iqaluit (Nunavut) "We elders are also to blame because we are not talking to young people as much anymore. We are relying on the teachers. Our parents did not have teachers or anyone else to rely on. [...] Some people think that because their children are able to speak English, even though they can't understand them, they follow them, because they think they are more capable than they are. That is not the way it is. [...] People fluctuate in the amount of time they spend talking to young people. Sometimes they become too preachy and other times they totally neglect them. This, too, is not good for young people."

Nunavut Inuit elder Itinnuaq

"We were advised from the time we were children to aim towards being good people. We have not done all that our ancestors did, but we have knowledge that our parents passed on to us. I think that if we were to impart this knowledge, it would really improve our young people's lives. I agree that we elders have not passed this knowledge on enough to our young people. If we started talking to them about what we know, I think the number of offences would go down. We seem to have been hiding our knowledge. We have based this on our thinking that it conflicted with [Christian] religion. We have a lot of knowledge that we should be passing on."

Parenting and counseling the young: voices from Nunavut expectations as to whom one is supposed to marry. While earlier studies mostly paid attention to formal aspects of kinship systems, more recent investigations have focused on the cultural notions that underlie how kinship and other social relations work in practice.

For example, in the everyday lives of Barrow and Kangersuatsiaq residents, the flexibility of Inuit kinship makes relatedness negotiable and almost entirely independent from biological links , while among contemporary Dolgan and Nganasan residents of the Taimyr Peninsula, relatives still play a major role in food distribution, sharing, and other aspects of everyday life.

The relevance of kinship has not diminished with urbanization. An illustration of this is provided in a case study from Greenland presented in the box below.

THE EXTENSION OF ARCTIC RURAL COMMUNITIES

Urbanization is a universal and well-studied trend. In the Arctic, where living off the land dictated low population densities and scattered settlements, it is a relatively recent phenomenon, associated with the rise of welfare state policies, industrialization, and the spread of wage employment.

The concentration of the population in permanent settlements took place during the twentieth century. Many of the major Arctic towns have been growing at a rapid pace, and they are becoming increasingly multi-cultural . The figure on the following page provides an illustration of the rapidity of urbanization in Greenland since 1950.

In the past few decades, increasing numbers of indigenous people are also settling down in larger centers away from their home areas. For instance, Oslo, Stǫckholm and Helsinki are playfully referred to as the largest Saami villages or siidas in the Nordic countries . According to a recent study , about 7,000 Greenlanders live in Denmark, which is equivalent to about 15% of the Greenlanders in Greenland.

Two thirds of them are women, and they are spread throughout the country rather than concentrated in the capital. In 2001, about 10% of Canadian Inuit lived outside the Arctic . Among the recently better-studied Native urban communities are the Yup'ik and Iñupiat in Anchorage, Alaska . In 2003, about 10% of the 274,000 inhabitants of Anchorage were Native or part Native people, which corresponds to almost 17% of the total Native population of the State .

According to Fienup-Riordan, outmigration to cities does not mean the severance of ties with home communities: "Yup'ik communities are not disintegrating, their lifeblood gradually seeping away. Many can be seen as actually In Greenland, family is important also for many people living in towns.

For example, in Nuuk where family members have to work many hours during the week, it becomes important for kinsmen to spend time together in their spare time, during holidays, family celebrations, and times of crisis. Also, because people live in a large-scale society, they have a need to stick together to handle outside crises. Voices gathered in research on the role of kinship among families in Upernavik, with 1,218 inhabitants, and Nuuk with 13,884 inhabitants illustrate this well: "A family is a person whom you can get support from, I don't know what I should do without my family. I need my family-it can be my own family or my husband's family." (35-year old woman).

"Generally most of the families keep together, but a lot of the time they have to work. Your kinsmen give you strength and help you through crises, it is therefore important that families keep together. If families don't stick together it will be harder for them to handle crises, because they will have no one to support them." (67-year old woman)

"I think families are more divided today maybe because they are so busy, but there are still families that are strongly connected. It is very important that families are connected [...] we teach our children to understand how important it is for us to keep together." (young mother, 35 years old) Although urban kinship can be utilized in many different ways, the basic ideology is centered around expectations about moral and mutual obligations in the realms of sharing, naming, and adoption, among others.

Relatives have many different kinds of mutual obligations in their lives and sharing defines the family. Family members strive to keep relatives together to avoid being isolated and having no one to support and share with. People have a choice as to whether or not to fulfill the mutual obligations of the family, and sharing does not always mean that kinship systems and forms of social organization are harmonic.

Gitte Tróndheim, University of Greenland, Nuuk Kinship in urban Greenland expanding and recreating themselves in unprecedented ways until today, when they are as strong and vital as at any time in their 2,500-year history. [...] Although geographically much farther-reaching than ever before, contemporary Yup'ik community continues to be characterized by wide sharing throughout the extended family, with money from wage labor and commercial fishing used to support a variety of harvesting activities to fulfill extended family needs" .

Also, Yup'ik identities are not abandoned, but reinforced in Anchorage, albeit in a modified out a study of the socio-economic dimensions of Yup'ik grass basketry.

During the first few years of the study, when Annie Don, my Yup'ik collaborator, and I traveled frequently throughout the Yukon-Kuskokwim Delta villages, we often arrived in communities with no prearranged place to stay. Almost everywhere, there would be an older person living alone who was willing to take us in.

Because it is unusual for Yup'ik / Inuit-especially older people-to live by themselves, I began to make inquiries. "Where's your daughter?" I'd ask. "Moved into Anchorage with the kids," would come the reply.

At the same time, I realized that Annie, who has lived in Anchorage since 1994, was part of this migration, as were her circle of mainly Yup'ik women friends. Furthermore, at the Alaska Federation of Natives Crafts Fair held in Anchorage each October, roughly 75% of the vendors are female.

So I wondered if the migration was evenly distributed between men and women. Sure enough, this demographic reflects the gender breakdown of Anchorage-based Alaska Natives as a whole. In 2000, there were slightly over 7,000 Alaska Native women 18 years and above residing in Anchorage, but only some 5,500 men.

How does urban in-migration affect Alaska Native women? For one thing, they confront on a daily basis political issues that would be at greater remove from them in the village. Some 50-75% of the urban-dwelling women support themselves through the sale of arts and crafts.

These commodities in turn depend for their appeal on exotic raw materials such as seal hide, walrus stomach, and basket grass only obtainable in rural areas. The question of whether urban Native people will have continued access to these products on public land is at present hotly debated and since their very livelihood depends on such access, they are beginning to participate more actively in the political process than would normally be the case. In 1998, the Alaska Federation of Natives, its collective patience worn thin by the absence of resolution, called for a public protest, summoning Alaska Natives to a march and rally in downtown Anchorage. For Alaska Natives generally-and Alaska Native women in particular-direct confrontation such as a protest march, represented a lapse from culturally approved styles of conflict-resolution, which tend to favor settling disputes by negotiation. Among her circle of friends, the question of whether to participate was hotly debated. So as I circled by Annie's house to pick her up, I wondered if she would act.

But there she was on the doorstep. "It's for my grandkids," she commented as she climbed into the car, "Anyway if I can't go home and pick grass, how will I make my baskets?" Her remark illustrates the urban and rural connectedness that characterizes so much about life for Alaska Natives today. form emphasizing the common Yup'ik heritage rather than the local origin.

Thus, outmigration extends the social network of homeland villages and gives rise to "translocal communities"-at least for a time. A similar case has been made for the Inuit in Montreal, where it appears that urban Inuit are using ethnic identity as an adaptive tool in a multi-ethnic city.

Due to the original vocation of anthropology as a "western" discipline striving to study "exotic others," corresponding studies of southern immigrant communities in the Arctic are much less abundant. For instance, Danish identities in Greenland are only recently attracting the attention of researchers, one of them a Greenlander.

The persistence of kinship and community network among Arctic emigrants to urban areas is similar to recent observations in urban Southeast Asia, Latin America, and Africa. So far, the pattern is quite different from the developments that took place in Europe and in North America in connection with industrialization and urbanization.

TREND SUMMARY

Most Arctic communities have experienced relentless and pervasive changes over the past decades. The disruption of communication between generations is as much a consequence as a cause of difficulties in social reproduction. Kinship ties and networks have not died out with urbanization but remain a focus for meaning and identity in the growing Arctic towns, and even in urban settings outside the Arctic homeland.

KEY CONCLUSIONS OF POLICY RELEVANCE

Arctic societies and cultures are highly adaptable and resilient and thus well-equipped for integrating change. The fact that they integrate modernity should be viewed positively rather than with nostalgia for traditions lost. The concept of traditions should be seen as a dynamic one: traditions do not and should not hinder development.

If there is a political lesson to be learned, it is that paternalistic attitudes, top down approaches, and change imposed from the outside are counter-productive, even if they seem to foster social and cultural perseverance in the short run.

In the long run, the transfer of decisionmaking powers from central to local authorities might be the most important factor in helping reduce social and cultural problems associated with rapid change. To reverse negative trends that have their origins decades or centuries ago, however, requires a lengthy "normalization" period. Thus, one cannot expect things to improve immediately.

In each Arctic region, there is a feeling of being distinct from regions further south, even where those regions are adjacent. Globalization increases worldwide connections, but the information and communication technology that fosters it is also used to intensify circumpolar connections.

The emergence of a pan-Arctic identity can be further encouraged. Finally, decision makers should not overestimate the predictive power of social science data.

The inherent complexity of social and cultural systems makes predictions that rest upon the extrapolation of particular factors or trends tentative at best. "Social planning" approaches of the past testify to these dangers. Only through further investments into the development of better datasets and methodological tool kits can these limitations be overcome.

GAPS IN KNOWLEDGE

Most Arctic states do not keep separate statistics for their Arctic regions regarding key indicators of social and cultural change. For example, it is virtually impossible to determine numbers of speakers of particular languages spoken in the Arctic or to find reliable data about the religious affiliation of Arctic residents. We therefore encourage the Arctic states and Permanent Participants to actively assist in the compilation and distribution of Arctic social science data.

The study of social and cultural processes in the Arctic has been dominated by the discipline of anthropology. While this has led to a number of excellent studies on the local level, regional and national data are much rarer.

The anthropological heritage of Arctic social science research has led to a particular emphasis on indigenous groups and peoples, at the expense of "mixed" and "newcomer" groups. Anthropological research in the former Soviet Union focused more strongly on processes of the past than in other Arctic countries. Thus, there is a lack of studies regarding contemporary trends in the Russian North, especially for the period between World War II and 1990.

The former Soviet Union is not alone in putting more emphasis on certain kinds of research than on others. Most Arctic regions have excellent data on one or another aspect of social and cultural processes at the expense of others. A circumpolar research agenda would overcome the limitations of these "national" research agendas and result in increased comparability of the acquired data.

While there is ample documentation of the effects of individual aspects of social and cultural change, there is little understanding of the cumulative effects of rapid change. This gap in knowledge adds to the weak predictive potential of current knowledge.

SUMMARY

Arctic societies and cultures have been undergoing changes since time immemorial, but the 20th century resulted in an unprecedented amount and speed of social and cultural change. Still, even periods of rapid change have not eradicated traditional Arctic social and cultural systems. Rather, the contemporary Arctic is characterized by various combinations of "indigenous" and "western" elements. Alternatively perceived as "loss" or "innovation," these social and cultural processes inevitably lead to new forms of mixing "old" and "new." It will be up to Arctic residents to determine which kind of mix will best serve their future needs.

The Arctic has "grown" considerably during recent decades. Not only have contacts and levels of cooperation increased within the circumpolar North-especially since the opening of the former Soviet North-but the interconnectedness of Arctic and non-Arctic communities has become more apparent. While the impact of southern power centers on Arctic communities has long been noticed, Arctic communities are gradually expanding their reach south and thereby carrying Arctic social and cultural traditions into regions far removed from tundras and northern forests.

Bibliography

Acquaah, George: *Understanding Biotechnology: An Integrated and Cyber-Based Approach*, NJ, Pearson/Prentice Hall, 2004.

Ausubel, F.M.: *Current Protocols in Molecular Biology*, New York, John Wiley and Sons, 1989.

Baker, Kenneth Frank: *Ecology of Soil-bourne Plant Pathogens, Prelude to Biological Control*, Berkeley, University of California Press, 1965.

Bauer MW: *Biotechnology-the Making of a Global Controversy*, Cambridge, Cambridge University Press, 2002.

Bauer MW: *Biotechnology-the Making of a Global Controversy*, Cambridge, Cambridge University Press, 2002.

Bernauer, Thomas: *Genes, Trade, and Regulation: The Seeds of Conflict in Food Biotechnology*, Princeton, Princeton University. 2003.

Bernauer, Thomas: *Genes, Trade, and Regulation: The Seeds of Conflict in Food Biotechnology*, Princeton, Princeton University. 2003.

Broach, J.R.: *The Molecular Biology of the Yeast*, Cold Spring Harbor, Cold Spring Harbor Laboratory, 1981.

Chrispeels, Maarten : *Plants, Genes and Crop Biotechnology*, Sudbury MA, Jones and Barlett Publishers, 2003.

Costanza, R.: *Ecological Economics*, Columbia Uni. Press., New York, 1991.

Daubenmire, F.: *Plants and Environment*, New York, Wiley, 1947.

Dusenbery, David B.: *Life at a Small Scale: the Behavior of Microbes*, New York, Scientific American Library, 1996.

E. Dereniak and D. Crowe, *Optical Radiation Detectord*, John Wiley and Sons, 1984.

E. Dereniak and D. Crowe, *Optical Radiation Detectord*, John Wiley and Sons, 1984.

F. Reif: *Fundamentals of Statistical Mechanics*, McGraw-Hill, Singapore, 1965.

Fransman M, Junne G, Roobeek A: *The Biotechnology Revolution?*, Oxford, Blackwell, 1995.

Fumento, Michael: *Bioevolution: How Biotechnology is Changing Our World*, San Francisco, Encounter Books, 2003.

Goodman, D.C.: *From Farming to Biotechnology: A Theory of Agro-industrial Development*, Oxford, Blackwell, 1987.

Jameson, J. L.: *Principles of Molecular Medicine*, Totowa, Humana Press, 1998.

Jameson, J. L.: *Principles of Molecular Medicine*, Totowa, Humana Press, 1998.

Li, Koibong and Leslie A. Lewis.: *A Novel Approach to the Evolution of Microorganisms*, New York, Vantage Press, 1987.

Lodish, H.: *Molecular Cell Biology*, New York, Freeman, 2000.

Lynch, J.M., and Hobbie, J.E.: *Micro-organisms in Action: Concepts and Applications in Microbial Ecology*, Blackwell Scientific Publications, Oxford, 1988.

Maarten, C.: *Plants, Genes and Crop Biotechnology*, Sudbury MA, Jones and Barlett Publishers, 2003.

Metting, Jr., F.B.: *Soil Microbial Ecology: Applications in Agricultural and Environmental Management*, Marcel Dekker, New York, 1992.

Metting, Jr., F.B.: *Soil Microbial Ecology: Applications in Agricultural and Environmental Management*, Marcel Dekker, New York, 1992.

Metting, Jr., F.B.: *Soil Microbial Ecology: Applications in Agricultural and Environmental Management*, Marcel Dekker, New York, 1992.

Raff, K. Roberts, and J. Watson, *Molecular Biology of the Cell*, New York: Garland Publishing, 1989.

Raff, K. Roberts, and J. Watson, *Molecular Biology of the Cell*, New York: Garland Publishing, 1989.

Rietman, Ed.: *Molecular Engineering of Nanosystems*, New York, Springer, 2001.

Sayler, G.S., Fox, R., and Blackburn, J.W.: *Environmental Biotechnology for Waste Treatment*, Plenum Press, New York, 1991.

Sheldon, J.K.: *Practical Environmental Bioremediation*, Lewis Publishers, Boca Raton, 1992.

Switzer, R.L.: *Experimental Biochemistry*, New York, W.H. Freeman and Company, 1977.

Teich, M.: *Fundamentals of Photonics*, New York: Wiley, 1991.

Wood, E.J. Ferguson: *Marine Microbial Ecology*, London, Chapman and Hall, 1965.

Yanoff, M.; Duker, J. S.: *Ophthalmology*, St. Louis, Mosby, 2004.

INDEX

D

E

F

G

H

I

L

M

N

O

P

❑❑❑